WindEurope Annual Event 2022

Journal of Physics: Conference Series Volume 2257

Bilbao, Spain
5-7 April 2022

Editor:

Amir R. Nejad

ISBN: 978-1-7138-5511-8
ISSN: 1742-6588

Printed from e-media with permission by:

Curran Associates, Inc.
57 Morehouse Lane
Red Hook, NY 12571

Some format issues inherent in the e-media version may also appear in this print version.

Institute of Physics
Dirac House, Temple Back
Bristol BS1 6BE UK

Phone: 44 1 17 929 7481
Fax: 44 1 17 920 0979

techtracking@iop.org

Additional copies of this publication are available from:

Curran Associates, Inc.
57 Morehouse Lane
Red Hook, NY 12571 USA
Phone: 845-758-0400
Fax: 845-758-2633
Email: curran@proceedings.com
Web: www.proceedings.com

TABLE OF CONTENTS

Journal of Physics: Conference Series 2257 (2022) 011001 doi:10.1088/1742-6596/2257/1/011001

PREFACE

This Special Issue presents selected contributions from the **WindEurope Annual Event 2022 Conference** that took place 5-7 April in Bilbao. A total of 15 papers are included, addressing one (or more) of the three conference themes:

- Onshore wind
- Offshore wind
- Electrifying the energy system.

Conference Committee for the Scientific Program

- Amir R. Nejad, Professor, Norwegian University of Science and Technology (NTNU)
- Stephan Barth, Managing Director, ForWind - Center for Wind Energy Research
- Athanasios Kolios, Professor, Strathclyde University
- Michael Muskulus, Professor, Norwegian University of Science and Technology (NTNU)
- Simon Watson, Professor, Delft University of Technology
- Olimpo Anaya-Lara, Professor, Strathclyde University
- James Carroll, Reader (Associate Professor), Strathclyde University

We thank all, more than 20, reviewers for their engagement and effort, without which these proceedings would not have been possible. As with previous WindEurope events, the European Academy of Wind Energy (EAWE) has been responsible for organizing the review process for scientific contributions and has contributed to the development of the conference sessions. WindEurope has been responsible for the organization and logistics of the conference, and we thank their highly professional staff and their associates for the excellent collaboration.

Proceedings editor

Amir R. Nejad, Professor, Norwegian University of Science and Technology (NTNU)

Journal of Physics: Conference Series 2257 (2022) 011002 doi:10.1088/1742-6596/2257/1/011002

Peer Review Statement

All papers published in this volume have been reviewed through processes administered by the Editors. Reviews were conducted by expert referees to the professional and scientific standards expected of a proceedings journal published by IOP Publishing.

- **Type of peer review:** Single Anonymous
- **Conference submission management system:** Morressier
- **Number of submissions received:** 15
- **Number of submissions sent for review:** 14
- **Number of submissions accepted:** 15
- **Acceptance Rate (Submissions Accepted / Submissions Received × 100):** 100
- **Average number of reviews per paper:** 2.2142857142857144
- **Total number of reviewers involved:** 21
- **Contact person for queries:**
  Name: Rébecca Brulé
  Email: rebecca.brule@windeurope.org
  Affiliation: WindEurope - Conference Team

Aerodynamic modelling of a novel vertical axis wind turbine concept

Laurence Morgan[1] **and William Leithead**[1]

[1] Dept. of Electronic and Electrical Engineering, University of Strathclyde, Glasgow, G1 1XW, UnitedKingdom

E-mail: laurence.morgan@strath.ac.uk

Abstract. This paper introduces the X-Rotor, a hybrid vertical-horizontal axis turbine concept designed to lower the cost of energy in the floating offshore environment. The development of a double multiple streamtube (DMS) simulation tool is presented alongside a thorough discussion of the secondary correction factors included in the model. New corrections for streamline curvature effects applicable to an airfoil where the blade normal plane is not aligned with the rotor plane are derived.

The DMS model is successfully validated against experimental data and against higher fidelity lifting line (LLT) simulations. Strong agreement is observed between the LLT simulations and the DMS simulations for both rotor averaged and azimuthally varying outputs, indicating that the DMS simulations can be used for future control simulations.

1. Introduction

Up to 80% of the global available resource for offshore wind is located in waters deeper than 60m and is therefore considered uneconomical for fixed bottom wind turbines[1]. This had led to large academic and commercial interest in floating offshore wind turbines (FOWTs), however, the cost of energy for floating offshore wind must be reduced to encourage widespread adoption. The use of Vertical axis wind turbines (VAWTs) has been proposed as an innovative solution to reducing costs for FOWTs, as VAWTs may be more suited to the floating environment due to multiple design synergies. These include a low drive-train position favourable for floating stability, and the compliance of floating support structures damping adverse cyclic loading associated with VAWTs [2]. This has led to resurgence in MW scale VAWT research, including the Deepwind project[3], INFLOW[4], seatwirl[5], SKWID[6], and the NOVA project[7]. A good overview of the current field is given in [8].

The X-Rotor represents a two-step innovation in the design of floating offshore wind, utilising a novel VAWT configuration combined with a secondary rotor power-take-off system. The primary rotor consists of a double V configuration designed to maximise the swept area, minimise the overturning moment typically associated with V-Rotors, and provide attachment points for the secondary rotors. The secondary rotors are utilised as a power-take-off mechanism in lieu of a conventional drive-train to circumvent both the torque ripple, and the low speed/high torque drive-trains that are inherent in conventional VAWT designs. A purely demonstrative representation of the turbine is shown in figure 1. A feasibility study [9] demonstrated that the X-Rotor has potential to reduce operations and maintenance costs by up to 55% and turbine capital costs by up to 32% yielding a cost of energy saving of up to 26%.

Figure 1. A non-technical rendering of the X-Rotor produced by Michael Egan.

The first step in effectively modelling the behaviour of the X-Rotor for control purposes is a fast, accurate model capable of simulating the primary rotor aerodynamics. Aerodynamic modelling of VAWTs is a less mature field than horizontal axis wind turbine (HAWT) modelling, and a number of different modelling paradigms exist including double multiple streamtube (DMS) models [10], actuator cylinder models[11] lifting line theory (LLT) models[12], panel models[13], actuator line models[14], and blade resolved CFD[7]. Whilst some cross code comparisons have shown that DMS models may not always perform accurately[15], other cross code comparisons have shown that they can effectively reproduce both rotor averaged and azimuthally varying results from higher order simulations [16], the experimental validation of DMS codes has also been largely successful [10]. Additionally, DMS models are computationally cheap, and relatively simple to implement. In this context, a bespoke DMS code has been developed at the University of Strathclyde to facilitate the simulation of vertical axis wind turbines with complex geometries and this code has been validated against experimental data and higher order simulations, the development and validation of this tool is presented in this paper.

This paper has the following structure; section 2 will provide an overview of the DMS model development, section 3 will test the validity of the model against experimental data and higher fidelity aerodynamic codes, finally section 4 will conclude the paper, outlining key findings and discussing future work.

The key novel contributions provided by this paper include; the derivation an angle of attack correction to account for streamline curvature in the case where the blade normal plane is not parallel to the rotor plane, a derivation of a conformal map to generate a virtual aerofoil that accounts for streamline curvature in the case where the blade normal plane isn't parallel to the rotor plane, the validation of a newly developed in house DMS code against lifting line theory results for a novel rotor shape using both rotor averaged and azimuthally varying quantities, and the publication of the power coefficients of the X-Rotor primary rotor.

2. DMS modelling

The DMS model was first introduced by Paraschivoiu in 1982 [17] which extended the multiple streamtube model of Strickland [18] to include two sets of actuator surfaces representing the upwind and downwind rotor halves respectively. In this implementation, 2D rotor disks are discretised into a number of streamtubes based on the rotor azimuthal coordinate with each tube crossing a pair of actuator surfaces representing the flow's interaction with the front and back rotor half. The basic implementation of Paraschivoiu was subsequently extended to describe streamtube expansion, however the streamtubes were still considered to be parallel, allowing for discretisation by the azimuthal coordinate [10].

Read and Sharpe proposed an alternative formulation of the DMS model which intrinsically accounted for the effect of streamtube expansion and the apparent deflection of the streamlines due to fluid momentum conservation, detailed in [19].

This paper follows the formulation of Sharpe, as it considers the basis to be more physically rigorous, and utilises the variable induction approach as model accuracy beyond revolution averaged variables is considered important. Here, an overview of the DMS method is presented, followed by an introduction to the relevant secondary aerodynamic phenomena and the engineering models available to model them.

2.1. DMS overview

In order to simulate a 3D turbine through DMS, the rotor must first be discretised vertically into a number of 2D segments parallel to the plane of rotation, as shown in figure 2*a*. Each 2D segment is then discretised into a number of streamtubes with the flow conditions in each streamtube characterised by the flow in the central streamline, as shown in figure 2*b*. Each streamtube crosses two actuator surfaces representing the interaction between the upwind and downwind rotor half, as shown in figure 2*c*. The flow through each streamtube can then be calculated using momentum balance equations under the assumption that the far wake from the upstream actuator surface can be taken as the inflow condition at the downwind actuator surface.

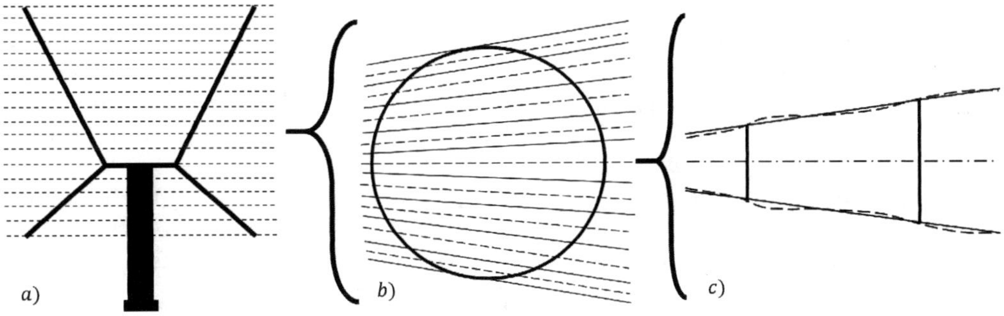

Figure 2. The discretisation procedure employed during DMS modelling.

The DMS model thus relies on the solution of momentum balance equations to resolve the local flow conditions. These are used to obtain the local blade forces which can, in turn, be integrated to obtain the rotor and blade variables of interest. This picture is complicated by the fact that the streamtube discretisation scheme cannot be completed with respect to any factor pertaining to the physical rotor geometry. This is because the location of streamtubes are not know a priori due to their dependence on the rotor loading distribution and the consequent streamtube expansion due to momentum conservation. To circumvent this issue, the rotor can be discretised by the angle subtended between the rotor radius vector and the incident streamline, henceforth referred to as the blade angle θ, rather than the angle of azimuth Θ. It should be noted that in the case of an unloaded rotor, the streamlines remain parallel and the blade angle reduces to the angle of azimuth ($\theta = \Theta$). Both θ and Θ are labeled in figure 3*a* for clarity.

Journal of Physics: Conference Series 2257 (2022) 012001 doi:10.1088/1742-6596/2257/1/012001

The solution procedure therefore involves three steps, the determination of the flow conditions at each blade angle (as calculated through momentum balance equations), the determination of streamtube geometry (as calculated through momentum conservation), and the integration of local blade forces into relevant rotor variables. These are each discussed in the subsections below.

2.2. Local flow calculations

At each actuator surface, the flow is reduced to a one dimension problem where the retardation of the flow through an actuator surface is characterised by the induction factor, a, with

$$U_{I[u/d]} = (1 - a_{[u/d]})U_{N[u/d]} \tag{1}$$

where U_I represents the windspeed at the actuator surface, U_N represents the wind-speed at the actuator surface if the surface was not present and the subscripts $[u/d]$ are introduced to specify if variables refer to the upwind or downwind actuator surfaces. Momentum theory dictates that the thrust coefficient at each actuator surface is given by

$$C_{T[u/d]} = 4a_{[u/d]}(1 - a_{[u/d]}). \tag{2}$$

This can be equated to the thrust coefficient as calculated though the non-dimensionalised, rotation averaged forces on the blade element projected onto the streamtube's central streamline, giving

$$C_{T[u/d]} = \left[sec(\theta)\frac{Nc}{2\pi R}\right]\left[\frac{U_{a[u/d]}}{U_{N[u/d]}}\right]^2 \left[C_n(\alpha_{[u/d]})cos(\theta)cos(\eta) - C_t(\alpha_{[u/d]})sin(\theta)\right]. \tag{3}$$

The first term represents the rotation averaging, with N representing the number of blades, c representing the local chord length, and R representing the rotor radius. The second term is an artifact of the de-dimensionalisation and represents the ratio between apparent velocity at the blade element, U_a and the incident velocity at the actuator surface in the no-load scenario, U_N. The final term represents the projection of the blade element force onto the streamline, with C_n and C_t representing the normal and tangential force coefficients, α representing the angle of attack, and η representing the angle between the blade normal plane and the rotor plane. The lift, drag, normal and tangential axes, as well as the associated blade angles are shown figure 3*b*.

At the upwind actuator surface, the no-load velocity is simply given by the free stream velocity ($U_{N[u]} = U_0$) whilst at the downwind actuator surface atmospheric pressure is assumed to have recovered with the no-load velocity given by

$$U_{N[d]} = (1 - 2a_{[u]})U_0. \tag{4}$$

The apparent velocity at the blade element is given by the sum of the incident wind-speed and rotation induced velocity

$$U_{a[u/d]} = \sqrt{[\omega R + U_{I[u/d]}cos(\theta)]^2 + [U_{I[u/d]}sin(\theta)cos(\eta)]^2}. \tag{5}$$

where ω represents the rotational speed of the blade element.

Finally, the angle of attack is given as the sum of the inflow angle, ϕ and the pitch offset, β

$$\alpha_{[u/d]} = \phi_{[u/d]} + \beta = \text{atan}\left(\frac{U_{I[u/d]}sin(\theta)cos(\eta)}{\omega R + U_I[u/d]cos(\theta)}\right) + \beta. \tag{6}$$

This scheme fully describes the flow in terms of two equations with two unknowns; the induction factors $[a_u, a_d]$. The solution to any streamtube can therefore be found by equating equations 2 & 3 at the upwind actuator surface to calculate $a_{[u]}$ then at the downwind actuator surface to calculate $a_{[d]}$.

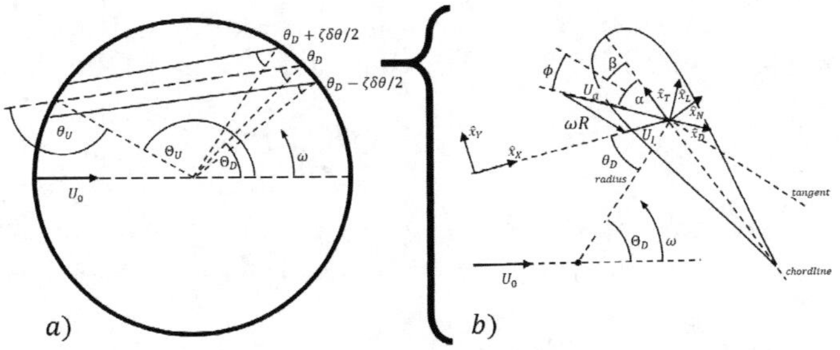

Figure 3. The streamline and blade element geometries involved in the DMS procedure.

2.3. Determination of streamtube geometry

The geometry of a single streamtube is shown in figure 3*a*, it has a central streamline which crosses the rotor at blade angles $(\theta_{[u]}, \theta_{[d]})$, and it's boundaries are given by the streamlines that pass at $(\theta_{[u]} \pm \zeta_{[u]}\delta\theta/2, \theta_{[d]} \mp \zeta_{[d]}\delta\theta/2)$.

The definition of a streamtube implies that the mass flow must be conserved which dictates that

$$U_{[u]}A_{[u]} = U_{[m]}A_{[m]} = U_{[d]}A_{[d]}, \tag{7}$$

where A represents the cross sectional area of the streamtube, and the subscript $[m]$ is introduced to represent the position at the at the mid-point between the surfaces. Assuming that the streamlines are straight within the region enclosed by the rotor segment, the area at the midpoint between the two actuator surfaces is equal to the mean streamtube area. The area of the upwind and downwind actuator surfaces can therefore be related to the mean streamtube area with

$$A_{[u]} = \left(\frac{2U_{[d]}}{U_{[u]} + U_{[d]}}\right) A_{[m]} = \underbrace{\left[\frac{2(1-2a_{[u]})(1-a_{[d]})}{(1-2a_{[u]})(1-a_{[d]}) + (1-a_{[u]})}\right]}_{\zeta_{[u]}} A_{[m]}, \tag{8}$$

$$A_{[d]} = \left(\frac{2U_{[u]}}{U_{[u]} + U_{[d]}}\right) A_{[m]} = \underbrace{\left[\frac{2(1-a_{[u]})}{(1-2a_{[u]})(1-a_{[d]}) + (1-a_{U})}\right]}_{\zeta_{D}} A_{[m]}, \tag{9}$$

where the square bracketed terms represent the upwind and downwind expansion factors, given by ζ_U and ζ_D respectively. To facilitate the solution of the problem, it is assumed that the midpoint streamtube area is given by

$$A_{[m]} = Rcos(\theta)\delta\theta, \tag{10}$$

which is independent of streamtube loading. With this, all variables are described and the geometry of any given streamtube can be determined.

As the central streamline is not deflected, the angle of azimuth corresponding to each streamline can be calculated though integrating outward from the central streamline:

$$\Theta_{[u]}(\theta) = \pi + \int_{\pi}^{\theta_{[u]}} \zeta_{[u]}(\tilde{\theta})d\tilde{\theta} \tag{11}$$

$$\Theta_{[d]}(\theta) = \int_{0}^{\theta_{[d]}} \zeta_{[d]}(\tilde{\theta})d\tilde{\theta}. \tag{12}$$

As the expansion factor is unknown prior before the streamtube is solved, cases using a cyclic pitch regime must employ a forward estimate technique to estimate the angle of azimuth

$$\Theta^{[i+1]} = \Theta^{[i]} + \zeta^{[i]}_{[u/d]}\delta\theta. \tag{13}$$

A key impact of the proper inclusion of streamtube expansion, is that the power extraction from the tandem actuator surfaces becomes coupled through the scaling of the actuator surface area.

2.4. Resolving and Integrating Blade Forces

With the geometry determined, the flow field is fully resolved, and all that remains is the integration of blade element forces to obtain rotor/blade variables of interest. The blade torque can be obtained by numerically integrating the elemental torque contribution over the blade span

$$Q_B(\Theta) = \int_0^L \frac{1}{2}\rho U_I^2 D_Q(\Theta) dl \tag{14}$$

where the local torque contribution is given by

$$D_Q = R[C_L sin(\phi) - C_D cos(\phi)] \pm (0.25 - \xi)c[\{C_L cos(\phi) + C_D sin(\phi)\}cos(\eta)]. \tag{15}$$

Here, the $\pm$ sign refers to the upwind and downwind case respectively and ξ represents the relative attachment position of the aerofoil (measured from the leading edge). Care must be taken in the evaluation of equation 14 to ensure that the integration is taken at a specific azimuth angle rather than at a blade angle as streamtube expansion is non-uniform over the rotor height. The rotor torque is found by summing the torque contributions from each blade with an appropriate phase offset

$$Q_R(\theta) = \sum_{i=1}^{N} Q_B(\Theta + i\frac{2\pi}{N}). \tag{16}$$

With the rotor torque obtained, the power is given by the product of the rotor speed and the integral of the torque over a single rotation

$$P = \omega \int_0^{2\pi} Q_R(\Theta) d\Theta. \tag{17}$$

Blade bending moments and other rotor force coefficients can be obtained through the integration and normalisation of the appropriate force vectors.

2.5. Additional aerodynamic phenomena

In the DMS method, the aerodynamic forces on the blade sections are obtained from blade polars; look-up tables that give the forces coefficients parallel and perpendicular to the direction of the flow. These polars are typically generated for a 2D wing in steady, rectilinear flow which is not representative of the blade element conditions for a vertical axis wind turbine. Additionally, the basic DMS method does not consider the flow entering the turbulent wake state. This section will give an overview of the unique phenomena associated with VAWT aerodynamics including; Dynamic stall, tip losses, streamline curvature, and the turbulent wake state, and will provide a description of the engineering models used to characterise these phenomena.

2.5.1. Dynamic Stall Dynamic stall occurs when an aerofoils unsteady motion lead to the angle of attack exceeding the static stall angle. It is typically characterised by an increase in the lift coefficient beyond the maximum static value at angles of attack exceeding the static stall angle as flow separation is delayed, followed by a sharp increase in drag coefficient and drop in the lift coefficient as the flow separates. During reattachment, lift coefficient remains below it's static value until the flow has reattached. A thorough discussion of the phenomena can be found in [20]. The two most common engineering models used to describe this phenomena in the context of VAWTs are the Gormont model and the Leishman-Beddows (LB) model, both adapted from helicopter aerodynamics.

The Gormont model is based on reproducing the forcing behaviour using an equivalent angle of attack, calculated based on Theodorsen's theory using the non-dimensional frequency of the dynamic motion, the angle of attack relative to the static stall angle, and the gamma function, obtained as a function of aerofoil geometry, non-dimensional frequency and Mach number. Originally applied to the transonic flows experienced by helicopter blades, the model was simplified by Strickland for incompressible flow with thicker ($t/c>0.12$) aerofoils, Strickland also limited the application of the model to regions where the angle of attack is greater than the static stall angle. Paraschivoiu proposed that the model should not be applied in the rear rotor half as stall was delayed due to the high free-stream turbulence and Masse recognised that the model often over-estimated the peak lift coefficient and introduced an interpolation scheme between the static and dynamic lift coefficients. A review of the Gormont model and it's modifications can be found in [21].

In general the Gormont model is considered to be simple to apply as all but one of the required constants can be readily obtained from the static blade polar data, with Masse's interpolation constant needing to be specified. A relationship between the chord length and the interpolation coefficient was put forward by Shires [22], however it should be noted that this relationship was only loosely validated, and has not been adopted in any future works. Additionally, the angle of attack history used to calculate the non-dimensional frequency can be generated using the free-stream velocity [19], so the dynamic stall model can be implemented directly into the DMS procedure.

The LB model represents a more complete unsteady aerodynamics model, with 3 distinct sub-models describing unsteady linear attached flow described using indicial response functions and approximations of Duhamel's integral, flow separation using Kirchoff-Helmholtz theory, and vortex shedding using a series of equations based on the Mach number and a number of empirical constants. For cyclical motion, the results from LB dynamic stall models typically take 2 to 3 rotations to converge [23]. This represents a challenge with respect to DMS methods, as the induction factor calculations also rely on iterative solution methods, however the challenge is typically overcome by ignoring the unsteady calculations in the calculation of the induction factors, then using the angle of attack history from the DMS simulation to re-calculate the blade loads using the dynamic stall model, as in [24]. A further issue in the implementation of LB type DMS methods is the issue of defining the empirical constants that require unsteady data for the aerofoil at the correction Reynolds number and Mach number. Whilst a range of data is presented in [25], and methods for generating the data using CFD simulations have been proposed in [26], the requirement for these constants and their sparsity in published work represents a barrier for the models implementation.

In this paper, the Gormont model with the modifications of Strickland and Masse are used in the description of dynamic stall.

2.5.2. Tip losses The aerodynamic force on a blade can be understood to be caused by the bound vorticity on the blade, this bound vorticity leaves the blade at the blade tips in the form of tip-vorticies, generating a complex structure behind the blade. This reduces the circulation at the blade tips, lowering the blade loads. This complex behaviour can be approximated using a tip-loss function, F. In the context of wind energy, the simplified tip loss model of Prandtl is typically invoked, using the assumption that the wake can be represented by a number of impermeable disks translated at the free-stream velocity. Numerous modifications of the tip-loss function exist for HAWTs and a thorough review is presented in [27].

With respect to VAWTs, these tip-loss functions are modified to maintain the same form of the tip-loss factor, but altering the disk spacing to better represent the vortex wake behind the rotor. The generalised form of the tip loss function is given by

$$f = \frac{2}{\pi} acos\left(exp\left[-\pi\frac{s}{d}\right]\right), \tag{18}$$

where s represents the distance from the blade tip, and d represents the distance between wake disks. Paraschivoiu proposes that the distance, d is given by

$$d = U_{w[u/d]}\frac{\pi}{N\omega}, \tag{19}$$

indicating that each of the N blades sheds a disk propagating at velocity U_W twice every rotation. Alternatively, Sharpe [19] proposed that the distance between the shed disks is given by

$$d = U_{w[u/d]}\frac{\pi R}{NU_0}. \tag{20}$$

These models are reconciled if the tip-speed ratio is equal to 1, however as the tip-speed ratio increases the tip loss function of Paraschivoiu estimates efficiency gains that are not realised in the case of Sharpe's correction.

The difference in these approaches is often masked by the normalisation scheme, applied to ensure that the tip-loss function is equal to unity (thus having no effect) at the blade center when considering H-Rotors. Paraschivoiu calculates the tip-loss contribution from the nearest blade tip, and normalises with respect to the tip-loss at the center, whilst Sharpe takes the product of both tip-loss functions, and normalised by the value of this product at the blade center. The fact that both approaches require normalisation at the blade center may imply that they may not be accurate for use in the case of V-Rotors or X-Rotors whereby the normalisation cannot occur as there is no position on the rotor blade where the tip-loss function is known to be equal to 1.

A further difference in approach can be found in how the tip-loss factor is applied. Here there are 3 options. Paraschivoiu uses the tip-loss function directly in the blade element calculations, retarding the flow velocity normal to the blade element by the calculated tip loss factor f, whilst Sharpe [19], indicates that the tip loss function should be applied such that the induction factor a, is replaced by the product af in the momentum theory calculations. In other works, such as that of [22], the approach used in HAWT BEM codes such as aerodyn[28] is applied, where the tip loss factor is applied such that

$$C_T = 4a(1-a)f. \tag{21}$$

The formulation used by Aerodyn is helpful, as it allows for Buhl's[29] correction for the turbulent wake state to be applied. The tool developed in this paper utilises the tip-loss function proposed by Sharpe, with correction applied as in equation 21.

2.5.3. Streamline curvature The effect of streamline curvature, also sometimes referred to as virtual incidence or virtual camber, describes the change in aerofoil behaviour between rectilinear and curvilinear flow. Typically, blade element polars are obtained from either wind tunnels or from simulation tools, both of which enforce rectilinear flow conditions. A blade section orbiting a central point sees curvilinear streamlines changing the local velocity and pressure distributions around the aerofoil and introducing a virtual camber and virtual incidence effects, an early study is given in [30]. Typically, a number approaches have been used to describe this effect. Muracia derived a relative scaling factor for the lift coefficient through integrating pressures around a flat plate[31]. Loth[32] introduced an angle of attack correction, almost identical to that which was later derived by Goude [33], recently Bangga[34] derived an angle of attack correction that re-defines the angle of attack from the blade attachment position to the blade leading edge. This reproduced the virtual incidence angle, even under pitched conditions, but did not reproduce the vertical camber effects. Sharpe[19] proposed a simplification of the angle of attack correction, whereby the relationship between the normal force coefficient and the angle of attack was considered linear, and the blade was assumed to be attached at the blade center, with this a correction was applied to the normal force coefficient.

Another approach that has been beneficial both for simulation and aerofoil design is the use of conformal mapping techniques. Applying a transform that converts the curvilinear streamlines experienced by the real aerofoil, into rectilinear streamlines, onto the aerofoil itself generates a virtual aerofoil that has the characteristics of the real aerofoil in those curvilinear conditions. The aerodynamic coefficients of this virtual aerofoil can then be obtained and used in the simulation, this was first pioneered by Migliore [30]. A review of the conformal mapping techniques applied to VAWTs is given in [35]. An issue with the use of these techniques applied to V-VAWTs is that the chord to radius ratio of the rotor changes significantly along the blade due to the changing rotor radius, requiring numerous virtual aerofoils to be generated for each blade. This significantly increases the time required to simulate new rotor designs, and it was therefore chosen to apply angle of attack corrections, such as those proposed by Loth and Goude.

One issue with the currently discussed correction factor is that each of them assume that the blade normal plane is parallel to the plane of rotation, which is not the case for V-Rotors or X-Rotors. Appendix A re-derives an angle of attack correction and a conformal map that can be applied to describe the effects of streamline curvature when the blade normal plane is not aligned with the plane of rotation. The angle of attack correction used by the tool developed in this paper is given by

$$\alpha' = \alpha + cos(\eta)\left\{\frac{1}{4}\frac{c}{R} + \frac{1}{2}\frac{(1-2\xi)c}{R}\right\}. \tag{22}$$

2.5.4. Turbulent wake state Experimentally, the parabolic relationship between induction factor and rotor thrust provided in equation 2 has been found to break down for induction values around (a >0.4) after which an approximately linear relationship emerges[28]. It is understood that this occurs because the wake behind the rotor disk enters a turbulent state and the relationship derived using actuator disk theory (based on Bernoulli's equation) breaks down. The classic correction applied in this state is obtained by taking a linear extrapolation of the thrust parabola, however this becomes problematic when a tip loss correction is applied to the thrust coefficient, leading to a discontinuity at the point of transition which can interfere with the iterative solution process. Bhul [29] proposed replacing the linear interpolation with a second order polynomial, which transitions smoothly from the momentum parabola to the maximum value at ($a = 1$). The value of the two endpoints, and the smoothness condition provide the 3 boundary conditions required to calculate the momentum coefficient.

A new development in actuator modelling in general was introduced by Ayati[36] who utilised an extension of Rankie-Froude momentum theory proposed by Sterios and Hultmark which is valid for a much wide range of loading, and doesn't predict wake reversal for highly loaded situations. This approach seems very promising for the application of DMS at higher tip-speed ratios where the rotor loading is generally above the range at which Rankie-Froude momentum theory is applicable.

The tool developed in this paper utilises the correction factor proposed by Bhul.

3. Model Validation

3.1. Validation against experimental data

Data from the Sandia 17m experimental wind turbine is chosen to validate the DMS model as the turbine was well instrumented, with blade force coefficients available [37] as well as the rotor power curve[38]. Additionally, because the rotor model is strutless, the uncertainties around the modelling of secondary rotor structures are minimised. The rotor was a 2 bladed straight-arc-straight Darrieus shape with a radius of 8.36m and a height to diameter ratio of 1 leading to a swept area of 187.1m^2. Each blade has a constant chord length of 0.61m with a NACA0015 section[38].

Figure 4 shows the experimentally determined rotor power curve against the output from the DMS model. In this case all of the previously described correction factors are active apart from the tip loss correction (as the rotor has no well defined blade tips). The agreement between the DMS model the experimental data in the area close to rated power is clear, and both the magnitude and location of the maximum power coefficient is well defined. At high tip-speed ratios the DMS model over predicts the available power, this may be because of an underestimation of the minimum drag coefficient of the blade at high tip-speed ratios. At very low tip-speed ratios, the model over predicts the power coefficient, this is due to the dynamic stall model over-estimating increase in the blade force associated with vortex lift.

Figures 5 & 6 show the normal and tangential force coefficients take at the equatorial rotor position for a low tip speed ratio ($\lambda = 2.39$) and an intermediate tip speed ratio ($\lambda = 4.6$) respectively. An initial observation is that the DMS model captures the experimental data relatively well. However, at the low tip speed ratio, it is clear that the dynamic stall model becomes active at an azimuthal angle of around 130° and causes an over-estimation in both the tangential and normal force coefficients after that point. A sharp drop in the tangential force coefficient is also erroneously predicted at 190° due to the model predicting a prolonged period before the flow has re-attached. In the intermediate tip speed ratio case, the force coefficients are relatively well described.

3.2. Validation against higher order simulation codes

Here, higher fidelity numerical simulations of the X-rotor are completed using the LLT codes CACTUS[39] and QBlade[12] and compared to the DMS model. The X-Rotor geometry is described in table 1. It should be noted that all values given at the blade and root tips can be interpolated linearly to obtain the span-wise distribution.

Initially, the LLT are are run without secondary corrections. As the models intrinsically account for the effects of the tip loss and the turbulent wake state, these are compared to a base DMS code which includes tip-loss corrections and the turbulent wake state. This examines how well models can reproduce each others behaviour in their most basic form, isolated from the differences cause by different dynamic stall and streamline curvature models. Subsequently, the CACTUS code is run with corrections for both streamline curvature, and a Leishman-Beddowes dynamic stall model and compared to the fully corrected DMS simulations.

Figures 7 & 8 show the comparison of the base model power curves and the corrected power curves. The general shape of the curve is well captured, and the tip speed ratio corresponding to the maximum power coefficient is in agreement between the 3 models. The DMS model over predicts the peak power coefficient relative to the other two models, however the relative uncertainty in the two LLT model outputs is of similar magnitude to the uncertainty between the DMS method and the CACTUS model. Comparing fully corrected DMS against CACTUS, there is considerable agreement in the low tip-speed ratio range, despite the fact that they use

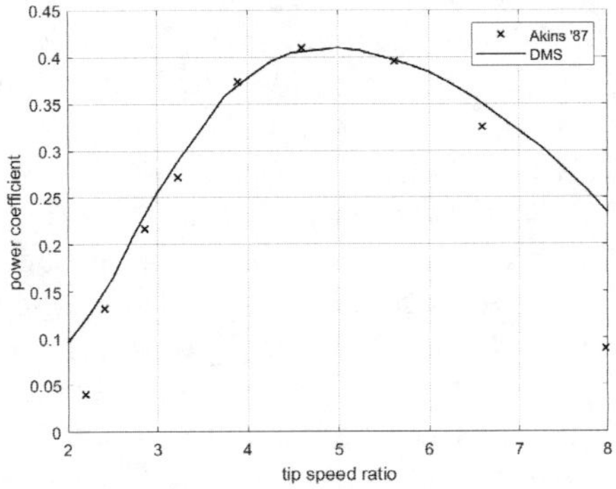

Figure 4. Power coefficient from the DMS model compared to the Sandia 17m demonstration turbine.

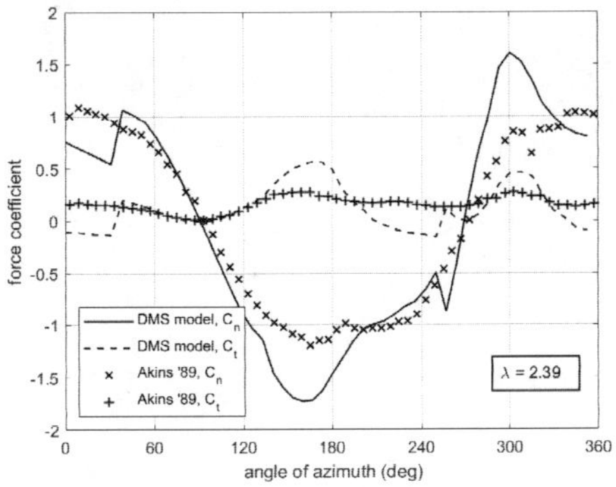

Figure 5. Normal and tangential force coefficients from the DMS model compared to the Sandia 17m demonstration turbine for $\lambda = 2.39$.

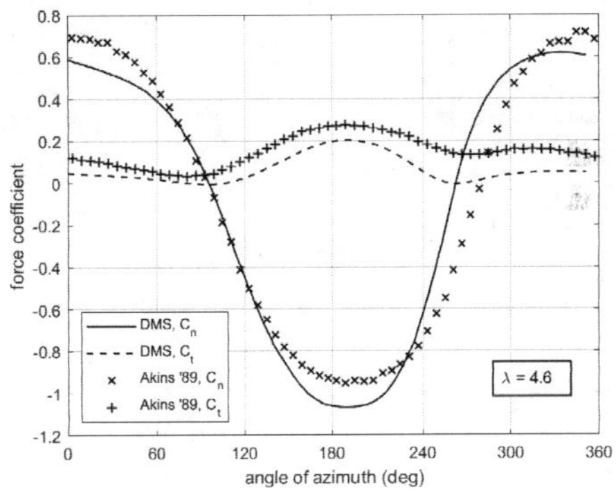

Figure 6. Normal and tangential force coefficients from the DMS model compared to the Sandia 17m demonstration turbine for $\lambda = 4.6$.

Table 1. X-Rotor Primary Rotor configuration.

Blade	R_{root}[m]	R_{tip}[m]	η [deg]	$\beta[deg]$	$c_{root}[m]$	$c_{tip}[m]$	Root section	Tip section
Upper	25	75	30	0	10	5	NACA0025	NACA0008
Lower	25	75	50	0	14	7	NACA0025	NACA0008

different dynamic stall models. The location and magnitude of the maximum power coefficient show good agreement, with the DMS model predicting a maximum power coefficient of 0.46 at a tip speed ratio of 4.25, whilst CACTUS predicts a maximum power coefficient of 0.47 and a tip speed ratio of 4.5. As the tip speed ratio increases further, the DMS model predicts a more rapid tailing off of the power coefficient compared to the LLT model. From the good agreement shown between both the base and fully corrected models, it is considered that the DMS model is able to produce power coefficient curves that are in agreement with higher fidelity lifting line codes.

Figures 9 & 10 show the torque coefficient obtained from the un-corrected models at a low tip speed ratio of $\lambda = 2$ and an intermediate tip speed ratio of $\lambda = 4.5$. There is strong agreement between the DMS method and the two LLT models for both tip-speed ratios, and the DMS method captures both the 4 peak torque spectrum at the low tip speed ratios, and the 2 peak torque spectrum at intermediate tip speed ratios. The amplitude of the torque coefficient is also well reproduced by the DMS model. Figures 11 & 12 show the torque coefficient obtained from the fully corrected models at a low tip speed ratio of $\lambda = 2$ and an intermediate tip speed ratio of $\lambda = 4.5$. Again, the agreement between the models is clear. At low tip-speed ratios, the shape of the torque curve is well described, however the peak torque coefficients are overestimated due to the difference in dynamic stall models. At intermediate tip-speed ratios, the torque coefficient curves and peak torque coefficients show good agreement. From this strong agreement, it can be inferred that the DMS model can accurately reproduce the torque profile as calculated using the higher order LLT simulation method, and can be used in future investigations into the behaviour of the X-Rotor.

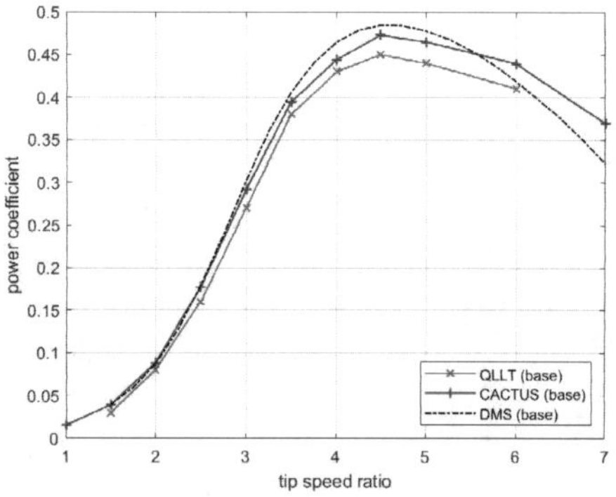

Figure 7. X-Rotor power curve without corrections

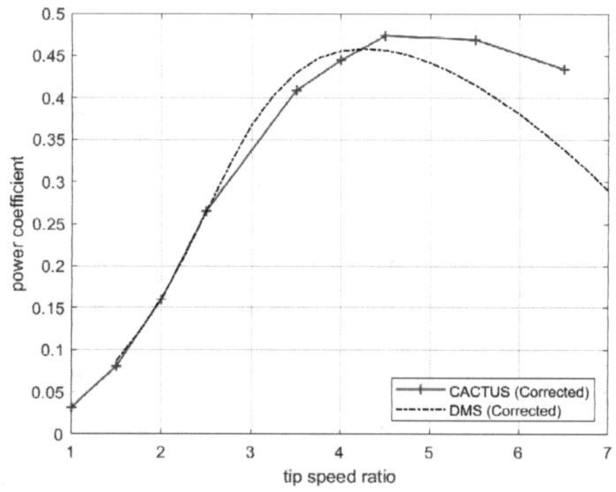

Figure 8. X-Rotor power curve with corrections

4. Conclusions and future work

This paper has described the development and validation of a fast and accurate DMS model for the goal of simulating the X-Rotor primary rotor. The underlying physics of the DMS method is rigorously introduced, and a thorough review of the secondary correction factors is introduced. The model is validated against both experimental data and higher fidelity simulation codes. The ability of the model to describe both rotor averaged and azimuthally varying values to a high degree of accuracy is demonstrated. Additionally, a derivation of flow curvature corrections in the case where the blade normal plane is not parallel to the rotor plane is provided in appendix A1 which, to the authors knowledge, has not been shown elsewhere.

With a simulation tool capable of accurately describing the X-rotor primary rotor aerodynamics, work is now ongoing into integrating this model into a wider control model that includes the secondary rotor aerodynamics and the structural and electrical dynamics of the system. Additionally, work is ongoing on the further validation of the DMS model.

5. Acknowledgements

The authors would like to acknowledge the EU H2020 XROTOR Project 101007135 and the EP/S023801/1 EPSRC Centre for Doctoral Training in Wind and Marine Energy Systems and Structures for funding this research.

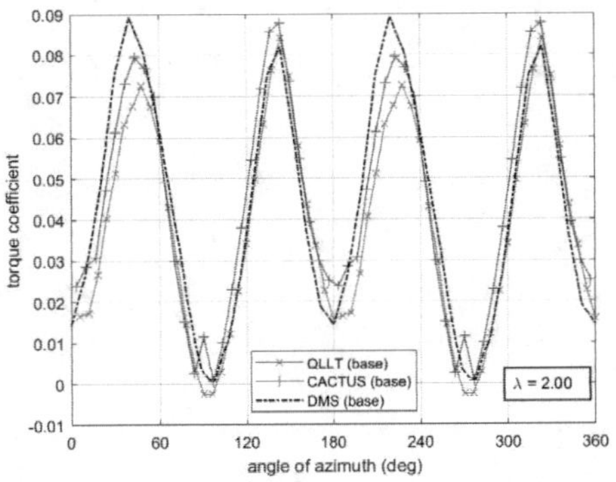

Figure 9. X-Rotor torque coefficient for $\lambda = 2$ without corrections.

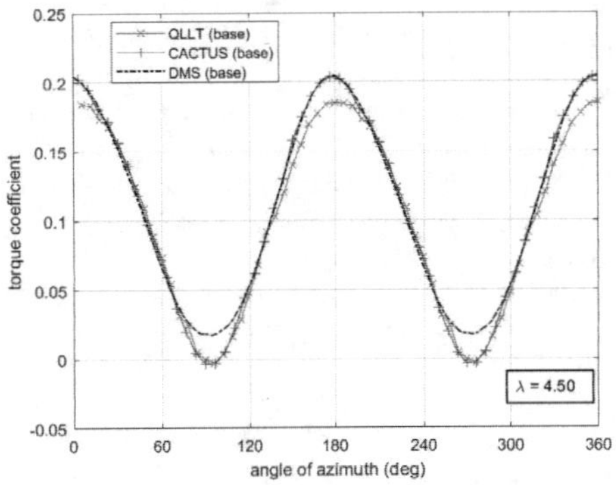

Figure 10. X-Rotor torque coefficient for $\lambda = 4.5$ without corrections.

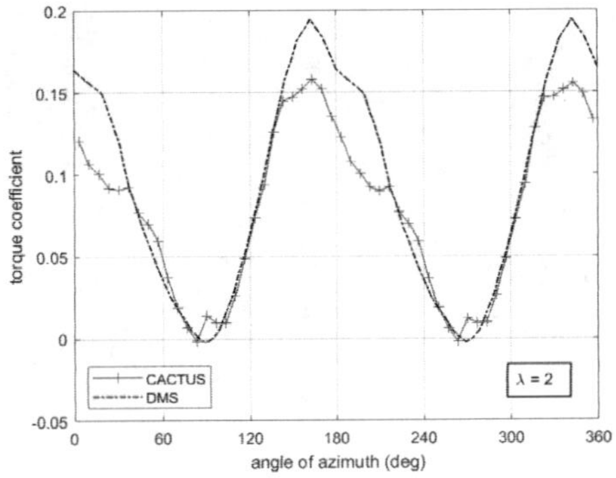

Figure 11. X-Rotor torque coefficient for $\lambda = 2$ with corrections.

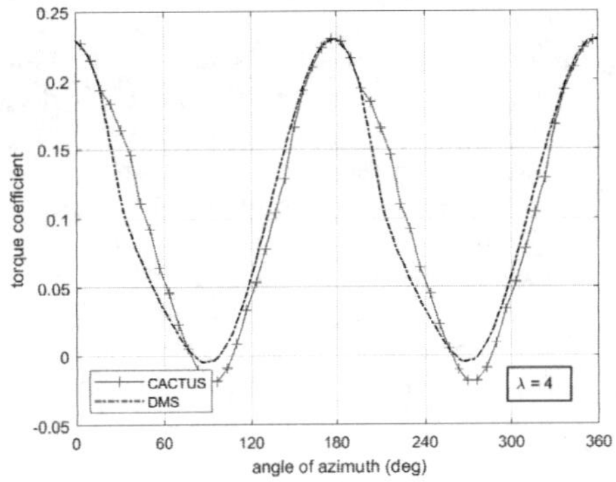

Figure 12. X-Rotor torque coefficient for $\lambda = 4.5$ with corrections.

References

[1] Wind Europe 2018 *Floating offshore wind energy a policy blueprint for Europe*
[2] Sutherland H J, Berg D E, and Ashwill T D 2012 *Technical Report for Sandia Nat. Labs.* SAND2012-0304
[3] Paulsen U S, Borg M, Madsen H A., Pedersen T F, Hattel J, Ritchie E, Ferreira C S., Svendsen H, Berthelsen P A, Smadja C 2015. *Energy Procedia* **80** 329-41
[4] INFLOW–Industrialization Setup of a Floating Offshore Wind Turbine *accessed 20/02/2022* Available online: *https://cordis.europa.eu/project/id/296043.*
[5] SeaTwirl *accessed 20/02/2022* Available online: *https://seatwirl.com/.*
[6] MODEC 2013 *Floating Wind & Current Hybrid Power Generation: Savonius Keel & Wind Turbine Darrieus* Product Brochure Tokyo
[7] Wang L, Kolios A, Delafin P L, Nishino T and Bird T 2015 *European Wind Energy Association Annual Conference and Exhibition 2015* Paris EWEA
[8] Hand B and Cashman A 2020 Sustainable Energy Technologies and Assessments **38** 100646.
[9] Leithead W et al 2019 J. Phys.: Conf. Ser.**1356** 012031
[10] Paraschivoiu I. 2002 *Wind Turbine Design: With Emphasis on Darrieus Concept* Presses Inter Polytechnique
[11] Madsen H Aa et al 2014 J. Phys.: Conf. Ser.**555** 012065
[12] Marten D, Pechlivanoglou G, Navid Nayeri C, and Oliver Paschereit C 2018. Journal of Fluids Engineering, **140** 021107
[13] Zanon A, Giannattasio P and Simão Ferreira C J, 2013 Wind Energy **16(5)** 661-80
[14] Mendoza V, Bachant P, Ferreira C, and Goude A. 2019 Wind Energy **22(2)** 171-88
[15] Ferreira C S, Madsen H A, Barone M, Roscher B, Deglaire P and Arduin I 2014 J. Phys.: Conf. Ser.**524** 012125
[16] Delafin P L, Nishino T, Kolios A and Wang L 2017 Renewable Energy **109** 564–75
[17] Paraschivoiu I. 1981 NASA. Lewis Research Center Wind Turbine Dyn
[18] Strickland J 1975 *Technical Report for Sandia Nat. Labs.* SAND-75-0431
[19] Freris L 1990 Wind energy conversion systems New York Prentice Hall
[20] Leishman G J, 2006 Principles of helicopter aerodynamics Cambridge Cambridge university press
[21] Masson C, Leclerc C, and Paraschivoiu I 1998 International Journal of Rotating Machinery **4(2)** 129–139
[22] Shires A. 2013 Energies **6(5)** 2501–2520
[23] Dyachuk E and Goude A 2015 Energies **8(2)**1353-72
[24] Wang K, Hansen M O L, and Moan T 2015 Wind Energy **18(1)** 91–110
[25] Dyachuk E, Goude A and Bernhoff H 2014 AIAA journal, **52(1)** 72-81.
[26] Hand B, Kelly G and Cashman A. 2021 Journal of Solar Energy Engineering **143(4)**: 044501
[27] Branlard E. 2011. Wind turbine tip-loss corrections: Review, implementation and investigation of new models *Masters Thesis* Risø DTU
[28] Moriarty P J and Hansen A C 2005 AeroDyn Theory Manual NREL NREL/TP-500-36881
[29] Buhl Jr M L 2005 A New Empirical Relationship between Thrust Coefficient and Induction Factor for the Turbulent Windmill State *Technical Report* NREL/TP-500-36834
[30] Migliore P and Wolfe W. 1979 17th Aerospace Sciences Meeting New Orleans 112
[31] Muraca R J, Stephens M V and Dagenhart R J 1975 Theoretical Performance of Cross-Wind Axis Turbines with Results for a Catenary Vertical Axis Configuration *NASA Technical Memorandum* NASA TM X-72662
[32] Loth J L and McCoy H. 1983 Journal of Energy, **7(4)** 313–8
[33] Goude A 2012 Fluid Mechanics of Vertical Axis Turbines *PhD Thesis* Uppsala university
[34] Bangga G, Dessoky A, Lutz T, and Krämer E. 2019 Energy **182** 673–88
[35] van der Horst S, van de Wiel J, Ferreira C S and García N R 2016. 34th Wind Energy Symposium San Diego California
[36] Ayati A A, Steiros K, Miller A M, Duvvuri S and Hultmark M 2019 Wind Energy Science, **4(4)**, 653–62
[37] Akins R E 1989 *Technical Report for Sandia Nat. Labs.* SAND-89-7051
[38] Akins R E, Berg D E, Cyrus W T 1987 *Technical Report for Sandia Nat. Labs.* SAND-86-2164
[39] Murray J and Barone M 2011 49th AIAA Aerospace Sciences Meeting, Orlando, Florida

Appendix A. Derivation of curvature correction model for V-Rotors.

Appendix A.1. Dynamics

The velocity at an arbitrary point on the surface of an aerofoil rotating around a central axis with the blade normal plane at an angle η to the rotational plane, as shown in figure A1a, is given by

$$\begin{bmatrix} U_x \\ U_y \end{bmatrix} = \omega \begin{bmatrix} R + ycos(\eta) \\ -xcos(\eta) \end{bmatrix} + U_0 \begin{bmatrix} sin(\theta) \\ cos(\theta)cos(\eta) \end{bmatrix} \tag{A.1}$$

the x, y axis with 0 at the blade attachment point as in figure A1b . In the case of a pitching aerofoil in rectilinear flow, shown in figure A1c, we have

$$\begin{bmatrix} U_x \\ U_y \end{bmatrix} = \tilde{\omega} \begin{bmatrix} y \\ -x \end{bmatrix} + \tilde{U_0} \begin{bmatrix} 1 \\ 0 \end{bmatrix} \tag{A.2}$$

Under the assumption of high tip-speed ratio ($\omega R >> U_0$) equation A.1 reduces to

$$\begin{bmatrix} U_x \\ U_y \end{bmatrix} = \omega cos(\eta) \begin{bmatrix} y \\ -x \end{bmatrix} + \omega R \begin{bmatrix} 1 \\ 0 \end{bmatrix}. \tag{A.3}$$

In this case, the similarity between the two motions is clear, with the equivalent pitch rate being equal to the orbital speed ($\tilde{\omega} = \omega cos(\eta)$) and the equivalent free stream velocity equal to the induced speed at the blade attachment point ($\tilde{U_0} = \omega R$).

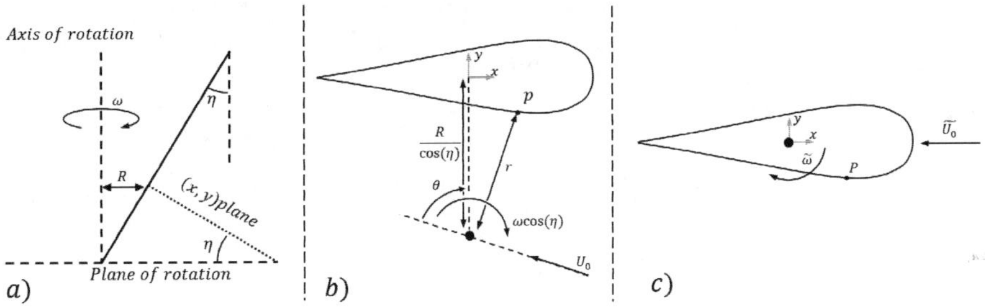

Figure A1. Similarity between the dynamics of a an aerofoil in orbital rotation and an aerofoil in a pitching motion.

Appendix A.2. Angle of attack Correction

An effective angle of attack correction can be obtained based on the quasi-static lift coefficient on a flat plate in a pitching motion given in it's general form by[20]

$$\alpha_{eff} = \alpha_G + \left[\frac{c}{2}\left(\frac{1}{2} - \xi\right)\frac{\dot{\alpha}}{U_0}\right], \tag{A.4}$$

substituting the effective pitch rate and effective freestream velocity we have

$$\alpha' = \alpha + cos(\eta)\left\{\frac{1}{4}\frac{c}{R} + \frac{1}{2}\frac{(1-2\xi)c}{R}\right\}. \tag{A.5}$$

Appendix A.3. Conformal mapping

Beginning with the simplified form of the velocity given in equation A.3, we can obtain the stream function, ψ which describes the shape of the streamlines and is given by

$$d\psi = U_x dy - U_y dx \tag{A.6}$$

integration yields

$$\psi = \left(\frac{\omega}{2}cos(\eta)y + \omega R\right) y + \frac{\omega}{2}cos(\eta)x^2 + C \tag{A.7}$$

Where C represents an arbitrary constant of integration. This can be rearranged with constants absorbed into the constant of integration and with the potential appropriately scaled to give

$$\psi = \left[\frac{y}{R} + \frac{1}{cos(\eta)}\right]^2 + \left[\frac{x}{R}\right]^2. \tag{A.8}$$

For a new coordinate system given by

$$x' = \frac{x}{R} \text{ and } y' = \frac{y}{R} + \frac{1}{cos(\eta)} \tag{A.9}$$

The streamfunction is representative of streamlines that are concentric circles around the origin. In the complex plane, with $z = x' + iy'$, the circles are represented by

$$z = \sqrt{\psi}e^{i\theta} \text{ for } [0 \leq \theta < 2\pi] \tag{A.10}$$

the mapping function

$$g(z) = u + i\nu = i|z|(1 + ln(z)) \tag{A.11}$$

transforms the concentric circular streamlines into straight rectilinear flow. The conformal mapping process then simply involves the conversion of the aerofoil coordinates into the coordinate system described in equation A.9 then the application of the transform given in equation A.11, and the transformation back into the traditional aerofoil coordinate system. Figure A2 shows an example of this transform applied to a NACA0025 aerofoil with a chord to radius ratio of 0.2 for $\eta = 45°$, compared to the mapping process proposed by Migliore in [30]. In this case, the additional incidence angle from the Migliore transform is 2.8°, whilst the added incidence angle from the new transform is 2°.

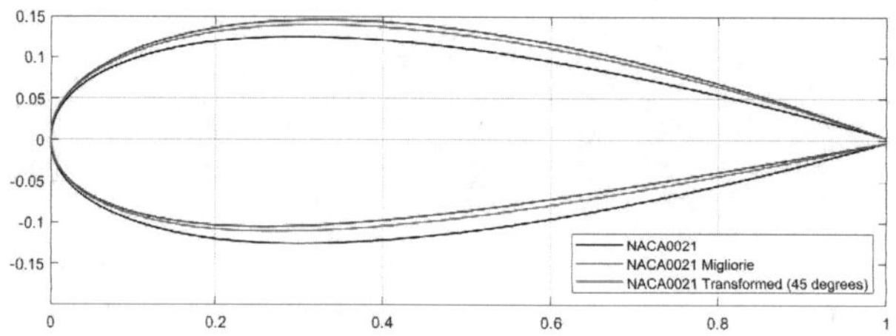

Figure A2. Demonstration of including the effects of the angle η on the conformal mapping process.

Development of a multi rotor floating offshore system based on vertical axis wind turbines

P Jamieson[1], C Simao Ferreira[2], P Dalhoff[3], S Störtenbecker[3], M Collu[1], E Salo[4], D McMillan[1], J McMorland[1], L Morgan[1], A Buck[1]

[1]University of Strathclyde, Glasgow, Scotland
[2]Delft University Technology, Delft, Netherlands
[3]Hamburg University of Applied Science, Hamburg, Germany
[4]Trios Renewables Ltd, Glasgow, Scotland

peter.jamieson@strath.ac.uk

Abstract The upscaling of wind turbines results in fewer units per installed MW reducing infrastructure and maintenance costs of offshore wind farms. Multi rotor systems (MRS), comprising many wind turbine rotors on a single support structure, are potentially a means to maximize the upscaling benefit in achieving larger unit capacities than is feasible or economic with the conventional, 3-bladed horizontal axis wind turbine (HAWT). The MRS has an inherent upscaling advantage which, for a system with many rotors compared to a single rotor, reduces the total weight and cost of rotor-nacelle assemblies by a large factor. An innovative MRS design is presented based on vertical axis wind turbine (VAWT) rotors of the 2-bladed, H-type. Many disadvantages of VAWT design compared to HAWT in a single rotor system (reduced power performance and higher drive train torque, for example) are resolved in the MRS configuration. In addition, reduced component number and simpler components is advantageous for reliability and O&M cost. This MRS concept has many synergies arising from the choice of VAWT rotors. Results comprise a high-level evaluation of system characteristics and the first stage of more detailed investigation of aerodynamics of the high aspect ratio VAWT.

1. Introduction

Upscaling of the unit capacity of wind turbines has been an ongoing trend in commercial wind turbine development. This upscaling continues due to the benefit, in offshore projects especially, in reducing the total number of maintenance sites and costs of foundations and electrical interconnections within a wind farm of given total capacity [1].

Multi rotor systems (MRS), comprising a number of rotors on a common support structure, are increasingly being considered [2], [3], [4] as an alternative to upscaling the standard 3 bladed, HAWT. The MRS has an inherent upscaling advantage which, compared to a single rotor system, reduces the total weight and cost of rotor-nacelle assemblies (RNA) by a very large factor [3], [5]. Obvious concerns regarding the MRS concept are in the aerodynamics of closely spaced rotors, the engineering and cost of the support structure, yawing of the system as a whole and maintenance of a system with a very large component count. Analytical [6], and numerical analyses based both on vortex methods and on computational fluid dynamics (CFD) [3], wind tunnel experimental results [7], [8] and field testing [9] confirm no power performance loss and varying gains with close lateral spacing of rotors, associated with a blockage effect. Ongoing work by HAW Hamburg [10], [11], [12] has considered the impact of rotor number and system layout on structure costs and other aspects of MRS design [13].

Diverse areas of MRS design have been addressed in PhD studies at the University of Strathclyde (UoS) including the overall system and loads [14], electrical systems [15], system yaw control using rotor thrust control [16]. In the Inwind.EU project, an availability and reliability analysis [3] highlighted the importance of having efficient methods to deal with minor faults. More detailed recent work on O&M highlights it as a critical area. Advantages of the MRS in avoiding the use of jack-up vessels through having in-built rotor handling systems, in redundancy of components reducing fault impact, in enhanced component reliability and the affordability of greater margins on fault critical components and in single machine faults affecting only a small fraction of system output are offset by the challenge of the larger total number of faults impacting on maintenance manpower demand and vessel logistics.

The distinctive aspect of the present work is in the development of an MRS system based on VAWTs rather than HAWTs. The idea is not entirely new, an MRS system based on spiral bladed VAWTs having been proposed by Coriolis Wind (no longer trading) [5]. The present system employs simple, 2-bladed H-type VAWTs aiming to show how limitations of such a design as a single rotor system maybe overcome with advantage in the MRS arrangement, exploring possible benefits in the avoidance of any additional engineering (as a floating system) for yawing in the sea and also likely benefits to O&M from reduced numbers of simpler components compared to an MRS based on HAWTs.

The overall aim is to develop a preliminary design of a floating VAWT based MRS and assess its potential to reduce cost of energy from offshore wind systems. The present work is the preliminary development of an outline design with a top-level assessment of some key design areas. The IEA 15 MW reference turbine [17], [18] is used as a baseline representative of a large offshore wind turbine design with public design information. This HAWT is compared with 48 MW MRS systems based on HAWT and VAWT rotors in the context of a large offshore wind farm. A multi-disciplinary team is assembled to address all relevant areas; fundamental aerodynamics, power performance and loads (TU Delft), structure and drive train (HAW Hamburg), floater (UoS) and O&M (UoS) using state of the art design tools. A detailed O&M model adapted for modelling MRS windfarms has been further developed in this project.

2. The MRS VAWT concept

2.1. Design characteristics of single VAWT types

Table 1 comparing VAWT types shows that the spiral and high aspect ratio VAWTS each solve one problem with the classic H-type but have their own downside as single rotor systems. The proposed integration of VAWT units into an MRS has great synergies that largely defeat the limitations of single VAWT design shown in Table 1.

Table 1. Comparison of single rotor VAWT designs

	Low aspect ratio VAWT	**Spiral bladed VAWT**	**High aspect ratio VAWT**
Power performance	*Cp* 5-10% less than for HAWT	*Cp* 5-10% less than for HAWT	Gain in reduced end loss offset by increased cross arm drag
Torque rating	Typically 2 to 3 times that of HAWT	Typically 2 to 3 times that of HAWT	*Equal or less than HAWT*
Torque cycle	Approx. sinusoidal	*Smooth*	Approximately sinusoidal
Rotor structure	Demanding cross arm joints	Efficient at small scale too complex for large	Blade support issue (many cross arms)
Support structure	Challenging	Challenging	Challenging (long cantilever)

2.2. The 48MW VAWT MRS

The design proposed (Figure 1) comprises 96 high aspect ratio VAWTs, each of 500 kW rated power, mounted on a single floating structure and resulting in a total system of 48 MW. The VAWT turbines are 2 -bladed units of the H type with 4 units in vertical columns coupled together and connected to a single generator. Direct drive PMG generators are primarily considered although there is a wide choice of power conversion options for MRS [15]. Turbine control is effected by generator reaction torque and routine braking is done electrodynamically with failsafe mechanical shaft brakes on each turbine unit. Salient design characteristics of the proposed VAWT MRS system are now considered.

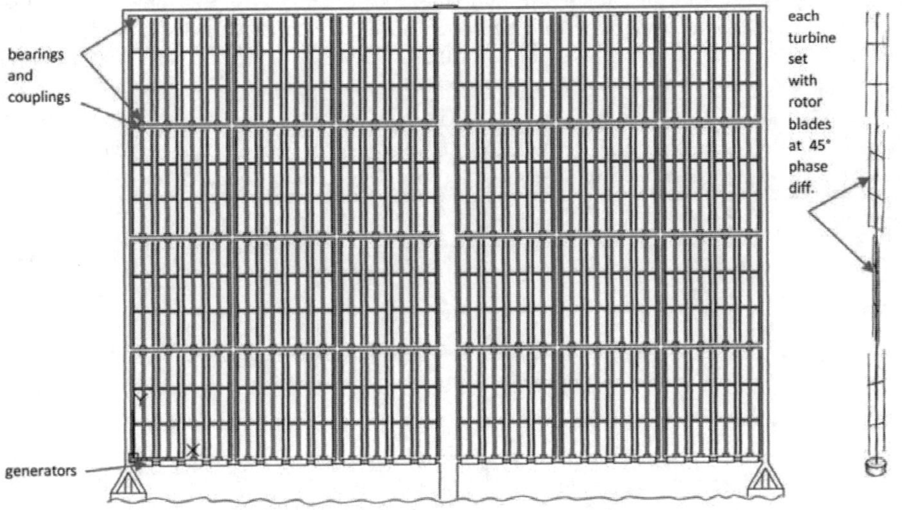

Figure 1. Multi rotor system of 48 MW based on high aspect ratio vertical axis wind turbines

<u>Structure packing</u> The limiting packing fraction of VAWTs of the H type is unity whereas for HAWTs it is minimum in triangular stacking at 0.907. This enables a reduction in the ratio of structure envelop area to total rotor swept area using VAWTs ~ 10%. The rectangular array is also much better suited for access and handling of components.

<u>Power performance</u> The high aspect ratio (5:1) minimises blade end losses. Blockage effects of close spaced rotors enhance power performance with added gain for counter rotating VAWTs.

<u>Rotor support</u> Structural support of a single high aspect ratio VAWT is challenging. Each MRS VAWT unit is integrated with top and bottom bearings avoiding cantilevering a long structure.

<u>Torque cycle</u> The 4 turbines in each column are rigidly connected with blade azimuth positions set differing in phase by 45° (Figure 1, right) giving a smooth, 8 phase torque input to the generator. This preserves simplicity and uniformity of rotor design.

<u>O&M impacts</u> Reliability is much enhanced avoiding pitch systems and yaw systems. In the VAWT MRS with rotors connected in sets of 4, electrical and control components are reduced by a factor of 4. O&M cost reduction also arises from the large unit capacity which is 3 times that of the largest commercial HAWT (Ming Yang MySE 16.0-242). The design intention is to avoid any need to return the complete system to shore base in the event of any RNA fault.

<u>Yawing</u> The design avoids the costs and complexity of a mechanical yaw system mounted on the floater or a turret with swivel bearing for yawing the floater in the sea. Instead, with the VAWT rotors being tolerant to varying wind directions, excepting ones closely aligned to the plane of the structure frame, the system can drift slowly downwind on conventional catenary moorings.

<u>Floater Torque reaction</u> With a large single floating VAWT, there is substantial additional demand on the moorings which must react the full generator torque. In the proposed MRS system, adjacent columns will counter-rotate. The net torque reaction on the floater will be very small.

3. Aerodynamics of high aspect ratio, H-type VAWTs in close spacing

3.1. Aspect ratio and blade end loss effects

VAWTs can in specific circumstances achieve high power performance [19] but gains are often limited due to support structure drag and tip loss. The rotor power performance coefficient, $C_p = 0.49$, of the IEA 15 MW reference wind turbine would rise by~ 8% to about 0.53 if the tip loss were eliminated. This suggests that there could be substantial benefit in a VAWT design that could much reduce tip loss (blade end) effects. In uniform flow, the C_p of a typical single H-type VAWT is a function of the aspect ratio and load distribution in the actuator cylinder's upwind and downwind portions. The multi rotor arrangement is based on a VAWT units with a 5:1 aspect ratio. From calculations based on a 3D actuator cylinder, [20], [21]. the influence of rotor aspect ratio on the maximum C_p (as a fraction of the C_p for infinite aspect ratio) for a VAWT primarily loaded upwind is illustrated in Figure 2. In this case the rotor has ¾ loading in the upwind half and C_p is expressed as a fraction of the value of C_p for infinite rotor aspect ratio. At the 5:1 aspect ratio selected for the VAWT MRS units, there is almost no loss associated with rotor aspect ratio effects.

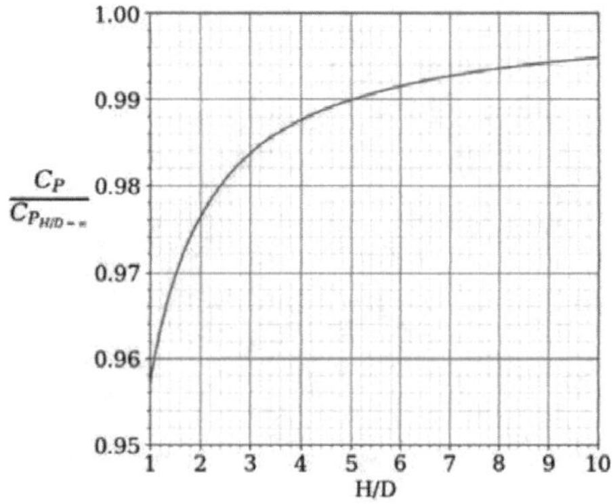

Figure 2. -Power coefficient of as function of aspect ratio, H/D.

3.2. Blockage effects at close spacing with co-rotation and counter-rotation

The Betz actuator disc limit applies when the flow has free expansion around the turbine. When this is not the case, more flow is forced through the turbine and more power may be extracted. Such blockage effects apply to tidal turbines confined in a narrow channel or in wind tunnels where the walls confine the total flow. A related effect termed 'local blockage' [22] obtains with rotors closely spaced in a plane where the flow expansion is inhibited by adjacent rotors. At optimal packing for HAWTs [22], the ideal maximum coefficient of power can be increased from the Betz limit 59.3% to the 79.8%. In the Innwind.EU project [3], a power performance gain of 8% of the 45 rotor MRS relative to a single HAWT rotor of the same total swept area was predicted. Similar effects of blockage apply to VAWTs [23]. Pairing VAWTs [24] side by side results in a power increase compared to two isolated VAWTs, and the clockwise-anticlockwise pair (also referred to as an 'inner-upwind' pair) provides the greatest increase in power; 13% -16% power gain was predicted [24] for a rotor spacing with swept areas 30% of diameter apart. In de Tavernier et al [25], it was shown that the proximity of the actuators leads to an increased thrust and power conversion, with gains of approximately 4% in power for thrust coefficients around 0.8. There is also evidence [26] that the inner-upwind combination produces a beneficial C_p characteristic with a flatter peak.

3.3. Rotor aerodynamic design

The VAWT proposed for the VAWT MRS is 2 bladed H-type using blade sections of the DU17DBD25/DU17W25 aerofoil [27] of length 88.7 m, diameter 17.73m and chord width 0.63 m. The performance is calculated using the actuator cylinder model originally proposed by Madsen [21] and implemented by Leblanc and Ferreira [28]. Rotor solidity, σ based on the definition $\sigma = Bc/R$ where B is blade number, c is chord width and R is rotor radius is approximately 14%.

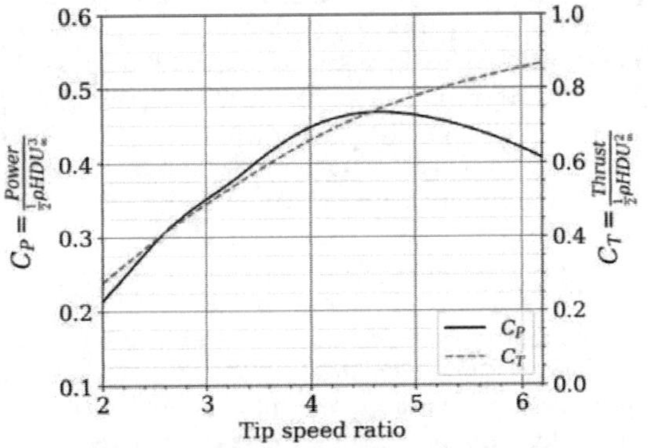

Figure 3. Power coefficient C_p and thrust coefficient C_t as a function of tip speed ratio.

Power and thrust coefficients as a function of tip speed ratio are presented in Figure 3. The predicted performance allows for cross arm drag but does not account for blockage benefit.

4. Rotor support structure frame design

4.1. Simplified space frame design

A simplified method for dimensioning of HAWT MRS space frames [29], [10] has been developed to design and compare different space frame topologies considering variations of geometric design parameters such as the space frame depth, space frame tower connection and the location of bearings.

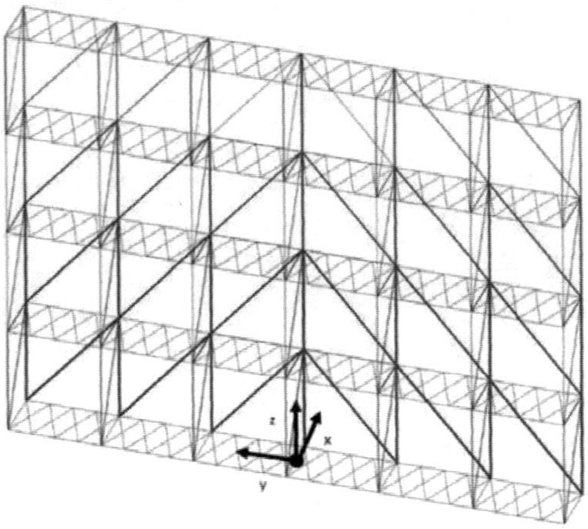

Figure 4. Space frame of the 48 MW VAWT MRS

The VAWT MRS space frame structure (Figure 4) has 4 rows and 6 columns with each structure cell holding 4 VAWT units, resulting in 96 VAWT units in total. A horizontal spacing between blade tips of 5% of the rotor diameter is used. The same spacing is used for the vertical distance between the rotors in each column. This leads to overall dimensions of 446 m width and 358 m height. The depth of the space frame (x-direction in Figure 4) can be varied in the concept analysis and is defined by a depth ratio: ratio of space frame depth to space frame width (dimension in y-direction in Figure 4). A truss structure with diagonal stiffening elements is used to allow a good load distribution.

4.2. Dimensioning process

The masses (10.76 t per RNA) of the 500 kW VAWT rotors are scaled from the 444 kW HAWT MRS rotors of the Innwind.EU 20 MW HAWT MRS [3]. Rated thrust force (85 kN per RNA) are similarly scaled. Each space frame member is dimensioned against yield strength and stability (Euler buckling) with applied safety factors. A thickness to diameter ratio of 1/120 is used. The space frame comprises 863 elements. There are around 220 distinct tube diameters after the initial dimensioning. The largest diameter is around 1.75 m, at the bottom row of the structure. A minimum diameter of 200 mm is set. To reduce the number of tube sizes, a classification into a set number of 20 bins is used. This increases the mass slightly but will benefit in having more standardised elements including joints and welds.

4.3. Space frame results

A space frame depth (Figure 5) of around 12% of the overall width (approximately 50 m) minimises space frame mass. The two points shown define a lower limit of possible depth ratio where the rotor swept area would touch the rear structural elements. To compensate for the simplified approach in the space frame dimensioning process, the space frame mass is doubled based on comparison with the more detailed evaluation in the Innwind project [3]. The resulting overall mass of space frame (5462.7 t) and RNA (1032.5 t) in combination with the centre of gravity is used in preliminary design of the floater.

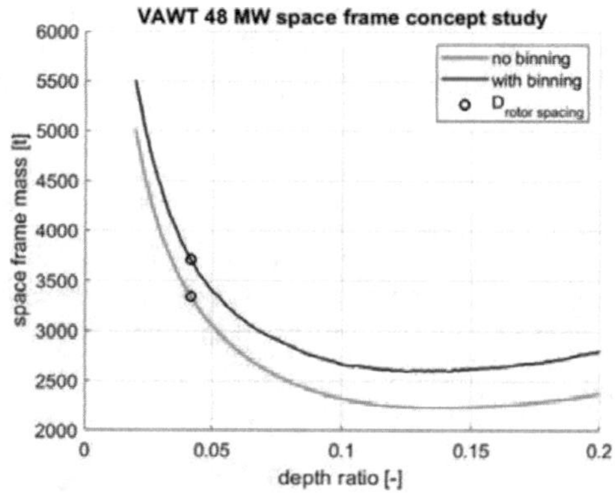

Figure 5. Space frame mass in relation to the space frame depth ratio

5. Drive train design

5.1. Drive Train Design

The shaft bearings and complete rotor assembly of each VAWT unit are supported in the space frame (Figure 1). The upper bearing is chosen to be the fixed bearing, (taking radial and axial forces) the lower bearing being the floating bearing (taking only radial forces). This leads to shaft tension by rotor gravity loads and avoids shaft buckling problems. In the proposed VAWT MRS design four VAWT units with aspect ratio 5 are connected via shaft couplings to a single generator at structure base The generator at base level allows for easiest maintenance access but involves the torsional wind-up of 4 shafts and puts a factor of 4 on shaft torque rating requiring a larger shaft in comparison to a solution with one generator per rotor. A generator at mid-level could preserve the torque smoothing and involve the wind-up of only two shafts. The most natural solution of one generator per rotor would lead to lightest shafts and additional control possibilities but negate the important O&M advantage of fewer electrical components.

5.2. Torque shaft design and drive train vibrations

Connecting 4 long shafts may lead to drive train torsional vibrations since each torque shaft in itself is an elastic torsion spring and each rotor comes with a substantial mass moment of inertia. The shaft is designed for static strength with a global safety factor of 2 and a static shear strength of 300 MPa. Only the lowest shaft must transmit the torque of all four rotors, but manufacturing is standardised using a common design for all VAWTs. At a fixed wall thickness ratio of $d_s/t_s = 60$ ($d_s$shaft diameter, t_s wall thickness) estimated diameters and wall thicknesses (Table 2) are relatively small in consequence of applying torsion-only loading. Comparing the shaft masses to HAWT of the same rated power, the VAWT shafts are heavier due to much greater length. The masses in Table 2 for HAWT MRS have been scaled from a 2.1 MW HAWT with a shaft mass of 8 tons in forged steel with a diameter of 700 mm at main bearing position [30].

Table 2. Dimensioning of shafts

Shaft property		IEA 15 MW	V 15 MW AR 1	V 15 MW AR 5	VMRS 2 MW	VMRS 0.5 MW	HAWT MRS
diameter	d_s [m]		2.1	1.6	0.47	0.3	
wall thickness	t_s [mm]		35.5	27.1	7.88	5.0	
weight	m_s [t]	153	409	535	7.9	3.1	0.9
Eigenfrequency 1st	ω [Hz]		0.0164	0.0214	0.0367	0.134	

After shafts being dimensioned, a torsional vibration analysis is performed. A mass moment of rotor inertia has been crudely estimated by assuming that 70% of the RNA mass is rotating in the rotor radius R.

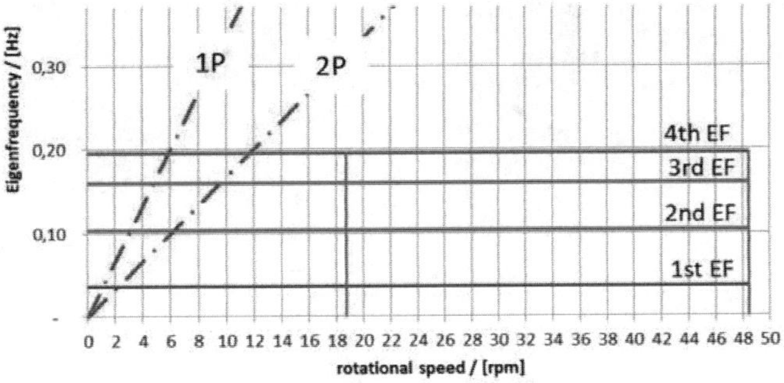

Figure 6. Drive train torsional vibration Campbell diagram for VAWT 4 rotor column

Eigenfrequencies for a drive train of n masses and n springs with the last spring end being fixed can be calculated as [31]. Results indicate a soft design with operation below 1P. The large cyclic torque variation of a single two bladed H-type rotor will in the 4-rotor column be averaged out at connection to the generator.

6. Floater design

Floating substructure types can be classified in A) ballast-stabilised, i.e. spar as in Hywind Scotland Pilot park B) waterplane area-stabilised i.e. barges and semisubmersibles (e.g. Windfloat by Principle Power), and C) mooring-stabilised, i.e. TLP, [32]. This initial assessment adopts a spar configuration for the simplest comparison to single rotor floaters although a semi-submersible is the probable future solution to avoid excessively deep draft. A parametric analysis of th VAWT MRS floating substructure based on scaling from [3], takes as input a mass above substructure of 14472.7 te, the centre of gravity vertical position at 156 m, the maximum aerodynamic thrust force as 8.164 MN and the height of 214.5 m for its point of application. The objective is the definition of the approx. geometric dimensions of the floating substructure and the amount of structural material required.

Key performance indicators are the static tilt inclination angle, θ, (calculated as in [32]) due to the aerodynamic inclining moment M_I of the VAWT MRS, and, as a cost indicator, the approximate mass of structural offshore steel necessary for the substructure. The spar, as represented by a simple cylinder with a constant radius R and a total draft T, has the vertical coordinate of the centre of buoyancy at $z_{CB} = -T/2$. The total substructure mass is the sum of its structural mass and ballast mass. The substructure structural (steel) mass is assumed to be 13% of the mass of displaced water [33], with a centre of gravity at $-T/2$, and the ballast mass can be calculated as the difference between the buoyancy force and the sum of the weight of the topside and the substructure structural mass, while its centre of gravity is calculated assuming a ballast tank at the bottom of the spar, with a ballast density equal to 1907 kg/m^3 [34]. Floatability is imposed by ensuring a ballast mass higher than zero.

The 5MW OC3 spar substructure steel mass [35] is considered as a benchmark for the steel mass. Feasible designs are in the coloured regions; white ones represent non-physical solutions (not floating or not statically stable). Reducing max θ which is usually in a range between 5° and 10° [36], requires a larger draft for a given radius, or a larger radius for a given draft and more costly substructure (Figure 7, right). This range allows for added dynamic response, due to wave oscillatory loads. Using the rated power divided by the substructure structural steel as a performance indicator, to be more efficient than the OC3 requires values (Figure 7, right) lower than 48MW/5MW = 9.6.

Considering a max θ of 10°, we can find floating substructure for the 48MW VAWT MRS with a better MW/tonnes ratio (i.e. lower than 9.6 in the graph) for deep draft (higher than 120m) and radii between 10 and 12.5m. The deep draft may be a limitation for this spar option, and a waterplane-area stabilised option will be investigated in future studies.

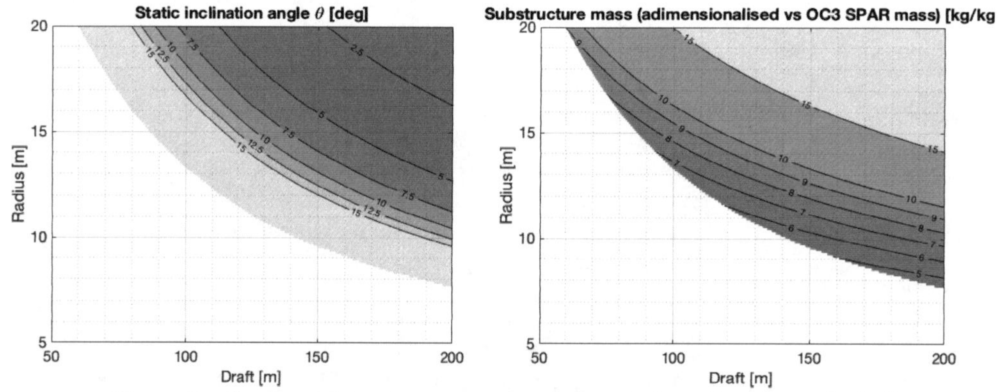

Figure 7. floating substructure parametric design: static inclination angle (left); dimensionless steel mass as a function of spar radius and draft (right)

The VAWT MRS, due to counter rotation of many rotor sets, imposes little torque reaction on the floater and its moorings, a significant advantage compared to large single turbine floating VAWT systems [37],[38].

7. O&M comparisons

7.1. Modelling of O&M

The O&M model is based on one originally developed by UoS [39] developed into the Trios OPEX model of Trios Renewables and extensively validated against the ECN (both analytic & time series versions) and MAINSYS (Shoreline) off-the-shelf OPEX tools [40], [41]. Iterations of this model have been used to support project decisions relating to, for example, the Beatrice Offshore Wind Farm (2016) and the SeaGreen project (2019). Trios produces a time series simulation representing wind farm operation on an hourly or half hourly time step, accounting failure and repair processes, allocating people and vessel resources as in actual operational practice over the operational life of a site. Met ocean and wind speed data are used in the model to constrain access and lifting operations.

7.2. Failure data

Table 3. Failure data groups

	HAWT		HAWT MRS		VAWT MRS		SH	MH	MV
	major	minor	major	minor	major	minor		number	
ELECTRICAL GROUP	0.623	2.472	0.063	0.303	0.034	0.259	3	83	22
MECH&STRUCTURE	0.431	3.302	0.011	0.177	0.004	0.121	3	59	23
YAW SYSTEM	0.007	0.182	0.014	0.364	0.000	0.000	3	1	0
PITCH SYSTEM	0.180	0.896	0.036	0.179	0.000	0.000	3	96	0
CONVERTERS	0.086	0.094	0.043	0.047	0.043	0.047	3	14	14

Without relevant failure data for VAWTs, failure rate data based on Carroll [42] reduced to the groups of Table 3 was used in evaluation of O&M comparisons. The absence of pitch and yaw systems in the VAWT MRS, the improved reliability of blades considering reduced surface erosion susceptibility of VAWTs at around half the tip speed of HAWTs, the improved reliability of generators and converters with uprating margins, and reductions in the number of components in various grouped categories is reflected in the data of Table 3.

7.3. O&M of Floating Wind Turbines

There is limited industry experience of floating offshore wind (FOW) technology with only 73 MW operational [43] but with much market growth expected. Crown Estate Scotland in January 2022 leased seabed options for approximately 16 GW of FOW. Inwater depth > 50m, the use of conventional major component replacement vessels will not be viable [44]. New solutions include floating-to-floating transfer, floating cranes, self-hoisting equipment and tow-to-shore [45]. The tow-to-shore strategy is the most favoured primarily using readily available, inexpensive tugboats and anchor handling vessels. However, longer weather windows of several days may be required for the disconnection and transit phases of the operation. The disconnection process of the mooring and electrical system considered sound in theory has yet to be tried in practice and allocated time and cost. There are also concerns regarding the availability of ports which could support such operations. In the VAWT MRS concept proposed, the system would never be returned to shore for any RNA failure. This reduces the requirement for long periods of access, which may be especially problematic with exploitation of sites further from shore with more challenging environmental conditions. For new 16 MW plus single turbines, the tow-to-shore strategy will be a lengthy process resulting in significantly large downtime.

7.4. Results

Wind farms ~ 1GW total rating single are compared as composed of 16 MW single turbines (represented by upscaling the IEA 15 MW wind turbine) and as composed of HAWT and VAWT MRS of 48MW rating.. Maintenance activities were conducted by a fleet of SOV (Service Operation Vessel) and/or SWATH (Small Waterplane Area Twin Hull) vessels.

Table 4. TRIOS OPEX Model simulation results

		VMRS	HAWT	HMRS
Annual Average	Energy Generated (MWh)	4.91E+06	4.71E+06	4.37E+06
	Energy Lost (MWh)	2.01E+05	3.00E+05	1.02E+06
	Revenue Generated (£)	£196,463,006	£188,279,283	£174,604,371
	Lost Revenue (£)	£8,036,340	£12,001,453	£40,780,819
	Capacity factor (%)	58%	56%	52%

Due to the small rotor size and built-in redundancy an MRS with many rotors, the criticality of immediate repair is eliminated. Single rotor faults compromise only a few percent of total output and none whatever if in some cases, redundant components can automatically take over to restore functionality. This has allowed a failure threshold to be introduced where transfer is only required when a predetermined number of failures has occurred. Accounting for the differing unit capacities of single turbine (16 MW), HAWT MRS (0.5 MW) and VAWT MRS (2MW, considering 4 0.5 MW units have a common generator) the modelling considered failures before visit of, 1, 10 and 2 for single HAWT, HAWT MRS and VAWT MRS respectively. The VAWT MRS (Table 4) is the best performing technology with respect to all metrics.

Despite the smaller rotor size and ability to implement a threshold, the HMRS experiences resource utilisation issues. With 96 rotors the HAWT MRS requires an impracticably large sized team and fleet to operate. The VMRS overcomes these issues due to its smaller number of rotors (24) and therefore reduced components requiring repair. This proved to be much more manageable with a fleet of only 3 vessels being required to maintain high availability for the site.

The VMRS is able to make use of the small rotors and batched failures, without becoming overwhelmed by the need for high resources for operation. Simulations show an impressive availability of 97%, which is very competitive against current turbines. It is recommended that the group/batch maintenance approach is taken for this concept as, while it takes a small hit in availability, this is outweighed by the reduction of the strain on staff and vessels, while also saving on vessel expenses such as fuel consumption.

8. Discussion and conclusions

It seems likely that in the MRS configuration, if not otherwise, the H type VAWT may match the power performance of large HAWTs. Outline structure design based on previous design experience of MRS structures does not flag any particularly problematic areas. Standardisation of RNA systems, and structure cells offers benefits in overall production cost and reliability beyond those associated with scaling. The drive train (shaft, bearings, brakes generators) are low torque aggregating to less than 20% that of competing large HAWTs leading to large savings in first cost, characteristic of MRS systems. However, coupling long shafts together leads to a torsionally-soft system and may leave concern about control response as yet unexplored although the torsional wind-up can be halved by having generators at mid-level. A check on floater design suggests the system should not be disadvantaged in mass and cost relative to its power rating. Recent studies have suggested that MRS with very many rotors can become problematic for O&M manpower and vessel resources and that O&M may be the most critical area for viability of the MRS concept, at least in systems with very many rotors ($>$ 40). The present analysis highlights the likely benefits to O&M of the VAWT MRS concept in eliminating some reliability critical components and having generally simpler and fewer components than HAWTs.

The MRS is well suited to hydrogen production [46] on site providing a large platform for the electrolyser, compressors and other necessary equipment. For electricity production, a low rated wind speed and low power density (332 W/m^2 for the IEA 15 MW design) minimises cost of energy considering the high electrical costs between each turbine output terminals and the land based grid. If production for hydrogen is on site and de-coupled from short term electricity demand, then the same total energy can be produced with a higher rated power but smaller turbines reducing the overall frontal area of an array such as in Figure 1 by ~40% and thereby much reducing system cost.

Acknowledgement The authors are pleased to acknowledge the support of the Engineering and Physical Sciences Research Council (EPSERC) to the Centre for Doctoral Training in Wind and Marine Energies of the University of Strathclyde under grant number EP/S023801/1.

9. References

[1] Valpy B. et al. 2017 *Future renewable energy costs: offshore wind* InnoEnergy
[2] de Vries E 2020 Utility cash brings full-scale 15MW Nezzy2 within reach *WindPower Monthly*
[3] Jamieson P et al 2015 *Innovative Turbine Concepts – Multi-rotor System* Innwind.EU Deliverable 1.33
[4] AltEnergyMag 2021 Wind Catching Systems launches floating offshore wind technology
[5] Jamieson P 2018 *Innovation in wind turbine design; 2nd Edition* (UK: Wiley and Sons Ltd.)
[6] Voutsinas S 2006 Vortex Methods in Aeronautics: How to make things work *Int. J. Computational Fluid Dynamics* **20(1)**
[7] Smulders P, Orbons S and Moes C 1984 Aerodynamic interaction between two wind rotors set next to each other in one plane *EWEC*
[8] Ransom D, Moore J and Heronemus-Pate M 2010 Performance of Wind Turbines in a Closely Spaced Array *Renewable Energy World* **2(3)** 32–6
[9] Weston D 2018 Measurable power gains found in multi-rotor Vestas concept *Wind Power Monthly*
[10] Störtenbecker S, Anselm R and Dalhoff P 2020 Simplified approach for the optimal number of rotors and support structure design of a multi rotor wind turbine system *J. Phys: Conf. Ser.* **1618** 032009
[11] Störtenbecker S 2021 *Fatigue analysis concepts for a multi-rotor support structure* Wind Energy Science Conference (WESC)
[12] Eichelmann T 2022 *Simulation von Windnachführungskonzepten für Multirotor-Windenergieanlagen (Yaw control simulation of multi-rotor wind turbine systems); Bachelor Thesis* (HAW Hamburg)
[13] Anselm R 2021 *Comparison of yaw control strategies of 20 MW multi-rotor systems* Wind Energy Science Conference (WESC)
[14] Jamieson P and Branney M 2012 Multi-Rotors; A solution to 20 MW and beyond? *Energy Procedia* **24** 52-59
[15] Pirrie P, Campos-Gaona D and Anaya-Lara O 2020 Comparison of electrical collection topologies for multi-rotor wind turbines. *Wind Energ. Sci.* **5** 1237-52
[16] MacMahon E, Stock A, Leithead W and Jamieson P 2015 *Yaw control for 20MW offshore multi rotor system* Proc. European Wind Energy Association
[17] Gaertner E et al. 2020 *Definition of the IEA 15-megawatt offshore reference wind* Golden Colorado National Renewable Energy Laboratory
[18] Allen C et. al. 2020 *Definition of the UMaine VolturnUS-S Reference Platform Developed for the IEA Wind 15-Megawatt Offshore Reference Wind Turbine* Golden Colorado National Renewable Energy Laboratory
[19] Ferreira C and Geurts B 2015 Aerofoil optimization for vertical-axis wind turbines *Wind Energy* **18** 1371–85
[20] Durand W 1935 *Aerodynamic Theory - Volume II Division E General Aerodynamic Theory* (Berlin: Springer).
[21] Madsen H 1983 *On the ideal and real energy conversion in a straight bladed vertical axis wind Turbine; PhD thesis* Aalborg University
[22] Nishino T and Draper S 2015 Local blockage effect for wind turbines *J Phys.: Conf Ser.* **625**
[23] Zanforlin S and Nishino T 2016 Fluid dynamic mechanisms of enhanced power generation by closely spaced vertical axis wind turbines *Renewable Energy* **99** 1213-26
[24] Vergaerde A *et al* 2020 Influence of the direction of rotation on the wake characteristics of closely spaced counter-rotating vertical-axis wind turbines *J. Phys.: Conf. Ser.* **1618**
[25] de Tavernier D, Ferreira C, Li A, Paulsen U and Madsen H 2018 Towards the understanding of vertical-axis wind turbines in double-rotor configuration *J. Phys. Conf. Ser.* **1037** 022015
[26] Vergaerde A *et al* 2019 Experimental validation of the power enhancement of a pair of vertical

axis wind turbines *Renewable Energy* **146** 181-187
[27] de Tavernier D, Ferreira C, Viré A, LeBlanc B, and Bernardy S 2021 Controlling dynamic stall using vortex generators on a wind turbine airfoil *Renewable Energy* **172** 1194–211
[28] Leblanc B, Ferreira C 2020 Experimental demonstration of thrust vectoring with a vertical axis wind turbine using normal load measurements. *J. Phys. Conf. Ser.* **1618(5)**
[29] Störtenbecker S, Dalhoff P, Tamang M, and Anselm R 2020 Simplified support structure design for multi-rotor wind turbine systems *Wind Energy Sci.* **5** 1121–8
[30] Dresig H, Holzweißig F 2016 *Maschinendynamik* (Springer)
[31] Herrmann J, Rauert T, Dalhoff P and Sander M 2016 Fatigue and fracture mechanical behaviour of a wind turbine rotor shaft made of cast iron and forged steel *Procedia Structural Integrity* **2** 2951-58
[32] Borg M and Collu M 2015 A comparison between the dynamics of horizontal and vertical axis offshore floating wind turbines *Philosophical Transactions of the Royal Society A: Mathematical, Physical and Engineering Sciences* **373(2035)**
[33] Anaya-Lara O, Tande J, Uhlen K and Merz K 2018 *Offshore Wind Energy Technology* John Wiley & Sons
[34] Leimeister M, Kolios A and Collu M, 2020 Development and verification of an aero-hydro servo-elastic coupled model of dynamics for FOWT, based on the MoWiT library *Energies* **13(8)** 1974
[35] Jonkman J 2010 *Definition of the Floating System for Phase IV of OC3* Golden Colorado National Renewable Energy Lab (NREL)
[36] Collu M and Borg M Design of floating offshore wind turbines In: Ng C and Ran L *Offshore wind farms: Technologies, Design and Operation* Woodhead Publishing 2016 359-85
[37] Cevasco D, Collu M, Rizzo C and Hall M 2018 On mooring line tension and fatigue prediction for offshore vertical axis wind turbines: a comparison of lumped-mass and quasi-static approaches *Wind Engineering* **42(2)** 97-107
[38] Borg M, Collu M and Kolios A 2014 Offshore floating vertical axis wind turbines, dynamics modelling state of the art. Part II: Mooring line and structural dynamics *Renewable and Sustainable Energy Reviews* **39** 1226–34
[39] Dinwoodie I and McMillan D 2013 Operational strategies for offshore wind turbines to mitigate failure rate uncertainty on operational costs and revenue *IET Renewable Power Generation* **8(4)**
[40] Dinwoodie I, Endrerud O, Hofmann M, Martin R and Sperstad I 2015 Reference cases for verification of operation and maintenance simulation models for offshore wind farms *Wind Engineering* **39 (1)** 1-14
[41] Sperstad I, Stålhane M, Dinwoodie I, Endrerud O, Martin R and Warner E 2017 Testing the robustness of optimal access vessel fleet selection for operation and maintenance of offshore wind farms *Ocean Engineering* **145**, 334-43
[42] Carroll J, McDonald A, McMillan D 2016 Failure rate, repair time and unscheduled O&M cost analysis of offshore wind turbines *Wind Energy* **19(6)** 1107-19
[43] Spearman et. al. 2020 *Phase II summary report; Floating Wind Joint Industry Project* CarbonTrust
[44] Ahn D, Shin S, Kim S, Kharoufi H and Kim H 2017 Comparative evaluation of different offshore wind turbine installation vessels for Korean west–south wind farm *International J. Naval Architecture and Ocean Engineering* **9** 45–54
[45] Bayati I and Efthimiou L 2021 *Challenges and opportunities of major maintenance for floating offshore wind* World Forum Offshore Wind e.V.
[46] Caine D et al. 2019 *Dolphyn Hydrogen phase 1 - final report*, ERM

Journal of Physics: Conference Series 2257 (2022) 012003 doi:10.1088/1742-6596/2257/1/012003

Development of *MOST*, a fast simulation model for optimisation of floating offshore wind turbines in *Simscape Multibody*

M. Sirigu, E. Faraggiana, A. Ghigo, G. Bracco

Department of Mechanical and Aerospace Engineering, Politecnico di Torino, 10129 Torino, Italy

massimo.sirigu@polito.it , emilio.faraggiana@polito.it , alberto.ghigo@polito.it , giovanni.bracco@polito.it

Abstract. the paper presents the development of an innovative non-linear, time domain numerical model for the simulation of offshore floating wind turbines, named MOST. The model is able to evaluate the movement of the platform in six degrees of freedom, the power production and the load cycles acting on the blades. MOST is implemented in Matlab-Simulink environment using Simscape Multibody. The aerodynamics is modelled with the blade element momentum theory and the hydrodynamics is modelled using WEC-Sim, a Simscape library developed by NREL and SANDIA. The use of Simscape offers great flexibility to quickly introduce complex dynamic systems such as hybrid wave-wind platforms, flexible platforms, or sea water active ballast systems. Additionally, Matlab provides useful toolboxes and extensive libraries for advanced control systems, linearisation analysis, parallelisation and generation of C code. The results of MOST are then compared to FAST, an open-source code widely used in the academic research. The case study is a 15 MW reference wind turbine installed on the Volturn US platform. The comparison shows a good agreement between the two codes with a significant reduction of the simulation time.

1. introduction

Floating offshore wind turbines represent a very promising technology in the field of renewable technologies: they allow the exploitation of vast areas of the sea at a great distance from the coast, characterized by more consistent wind speed and availability, reducing visibility from the coast, and increasing electrical productivity. The traditional depth limits imposed by fixed structures are overcome, allowing installations in deep seas such as the Atlantic Ocean and the Mediterranean Sea [1].

Reliable and fast numerical models for simulating the system are essential to ensure optimal platform design. These models must be able to compute the movements of the platform, the mechanical stresses acting on the components and to estimate energy productivity with the required accuracy.

FAST, an open-source software developed by NREL, represents the state of the art for offshore wind turbines simulations [2]. Numerous experiments have been conducted to validate the FAST code [3]–[6]. However, Fast is difficult to access, as well as complex to implement and modify.

This article presents an innovative non-linear time domain model written in Matlab-Simulink environment, named MOST (Matlab for Floating Offshore wind turbine). The model computes the motion of the platform, the power production, and the mechanical stresses acting on the blade root. The results are then compared with FAST to verify the accuracy and simulation time.

The reason to develop a new alternative model compared to similar open-source programs derives from the need to have a user-friendly, flexible and multifunctional model: thanks to the extensive libraries in Simulink, it is possible, for example, to integrate the platform with the dynamics of wave energy converters (WEC), to simulate an active control of the water ballast, to simulate flexible platforms such as *Spiderfloat* [7], or to simulate a platform with two turbines.

The Simscape Multibody library was used in this context, which easily allows to build a dynamic model in six degrees of freedom.

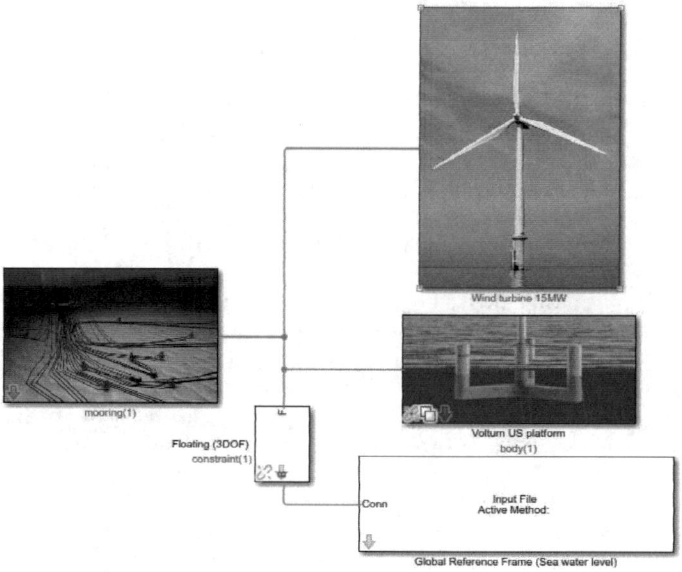

***Figure 1**: numerical model in Simulink environment*

2. Numerical model development

The numerical model includes:

1. Dynamic module, that computes the motion of the platform in 3 or 6 degrees of freedom and the total forces acting on the blades, tower, shaft and bearings including the aerodynamic and hydrodynamic forces, inertial, centrifugal and gyroscopic forces.
2. Aerodynamic module, that computes the aerodynamic forces using look-up tables of precalculated values that follow the blade element momentum theory.

3. Hydrodynamic module, that is computed using Wec-sim [8], a Simscape library developed by the collaboration between NREL and SANDIA.
4. Mooring module, that includes two options: a quasi-static mooring matrix that considers the non-linear stiffness of the chains and Moordyn that is included in Wec-sim library.
5. Control system module, that evaluates the blade pitch and generator torque response; The default control algorithms implemented are the baseline control and ROSCO control.

2.1 Aerodynamics

The aerodynamic forces are calculated using the blade element momentum theory as described in the FAST's aerodyn15 module, which represents the standard model of aerodynamic modelling. Unlike FAST, the aerodynamic forces are computed using a look-up table. The contribution of each blade to the axial and tangential forces and moments is a function of three input variables: the average wind speed on the blade, the rotor angular speed and the blade pitch. In order to decrease the points necessary for discretization, the angular speed and the blade pitch are discretized around the steady state points, thus excluding the points that are never reached in operating conditions. The number of points is evenly spaced.

Discretisation parameters	Wind speed	Rotor speed error respect to steady state	Blade pitch error respect to steady state
Interval	3÷25 m/s	± 1.5 RPM	± 12 deg
Number of points	45	5	25

***Table 1**: discretisation parameters for the aerodynamic forces*

The steady state values of a same turbine can vary with different types of control. In this case, the steady state values are taken from FAST simulations implemented with the baseline controller as described in [9]. The simulation is realised with a fixed tower and the precone and tilt angle of the rotor are null. The wind is steady and there is no shear effect. The gravity acceleration is set to zero to exclude the gravitational forces. The axial and tangential forces and moments are taken from the outputs of FAST, specifically RootFxc1, RootFyc1, RootMyc1, RootMxc1. The simulation lasts 90 seconds to overcome the transient effects. The last value of simulation is taken as steady state point. The rotor speed and blade pitch are obtained for each wind speed.

The points outside the steady state values are taken by disabling the rotor degree of freedom and manually setting the rotor speed and the blade pitch. The figures (3) summarize the trends of torque and axial thrust acting on the entire rotor considering an average wind speed.

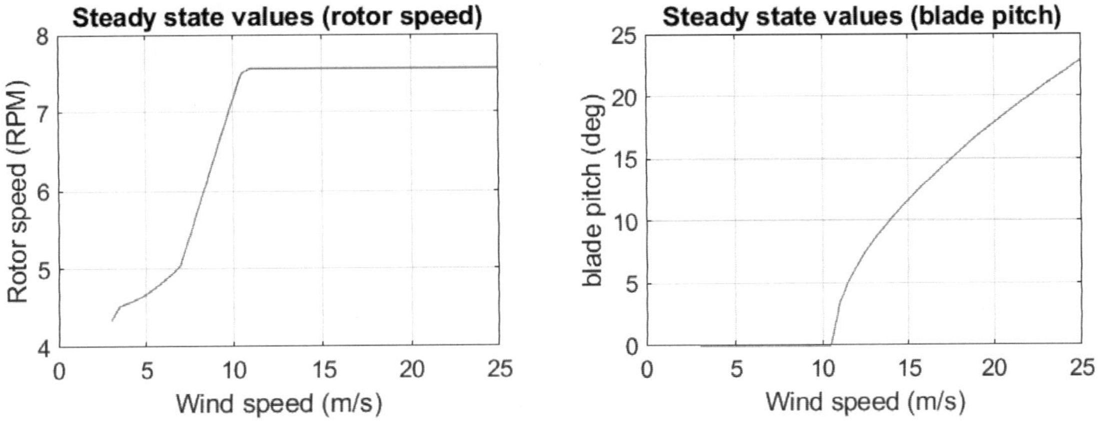

***Figure 2**: rotor speed and blade pitch steady state values from FAST*

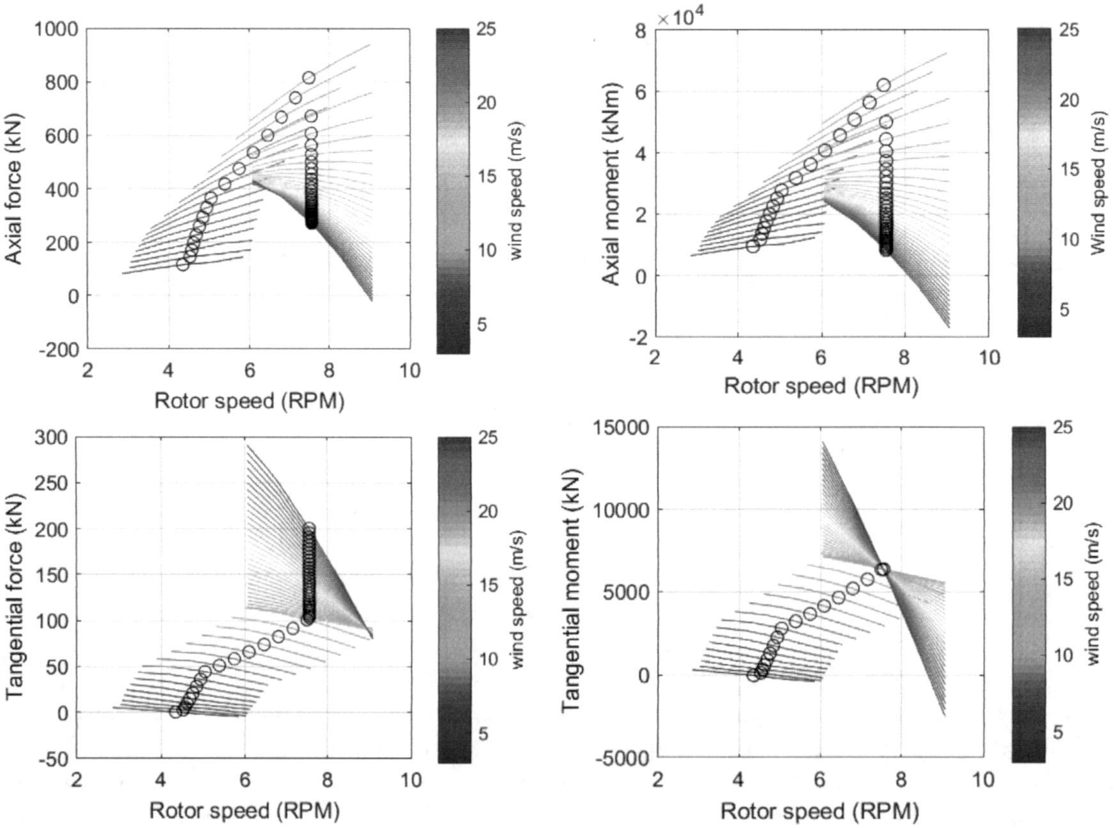

Figures 3: *Trend of the axial and tangential forces and moments for IEA 15 MW reference wind turbine as a function of rotor speed and wind speed. The blade pitch corresponds to its steady state value for each wind speed. The black circles represent the values of the forces in steady state conditions.*

The axial and tangential forces and moments are applied at the blade root. As expected, when the rotor reaches the rated speed (7.56 RPM), the tangential moments (that gives the torque to the electric generator) above the rated wind speed (10.59 m/s) converge to the same value of the torque, giving the rated power output. The slope of the tangential moment changes with the wind speed because the power to pitch sensitivity increases with the blade pitch, as described in paragraph 1.3. The axial force (that gives the thrust on the rotor) and moment decrease when the blade pitch increases.

The wind speed is calculated with the NREL Turbsim software [10]. The wind is represented as a two-dimensional grid 270 m x 270 m with a discretization of 17 x 17 points. The mean wind speed refers to the speed at 100 meters above the sea water level. The wind profile is a power law with an exponent of 0.14.

The average wind speed is determined by interpolating four points for each blade in the wind grid along the blade length, specifically at 50, 70, 90, 110 meters. The reason of a discretisation starting from the middle of the blade and not from the root is that the wind speed has more influence at the final section of the blade, and indeed it gives a better adaptation to the FAST results. The horizontal hub speed, due to surge and pitch oscillation, is added to the wind speed.

The aerodynamic forces do not take into account the flexibility of the blade and tower (rigid body assumption), the deflection of the wake due to the rotor misalignment with respect to the wind, and the wake dynamics.

2.2 Hydrodynamics and mooring

The hydrodynamics of the floating structure is computed using WEC-Sim [4]. WEC-Sim solves the dynamics of the floating bodies in time domain from a frequency-domain boundary element method. The hydrodynamic properties (linear hydrostatic, added mass, radiation damping, and wave excitation coefficients) can be evaluated with external softwares like Wamit, Ansys Aqwa, Nemoh, or Capytaine. The results of this tool have been validated in previous articles [11], [12].

Two options are implemented for the mooring: the first one consists of quasi-static mooring look-up tables that have been implemented within the Simulink model to account for the non-linear stiffness of the chains by solving the equations of the catenary [13]. The second one is Moordyn, a WEC-Sim built-in module in which the contributions of inertia and the viscous friction of the chains are considered.

2.3 Control

the control system is a baseline control, that is widely used in the wind industry. An in-depth analysis of this type of control can be found in [9]. The implementation of the baseline control for the 15 MW reference wind turbine is explained in this paragraph. The generator torque behaviour is described in the figure (4). It follows the optimal torque for the maximisation of the power extraction and depends only on the generator speed.

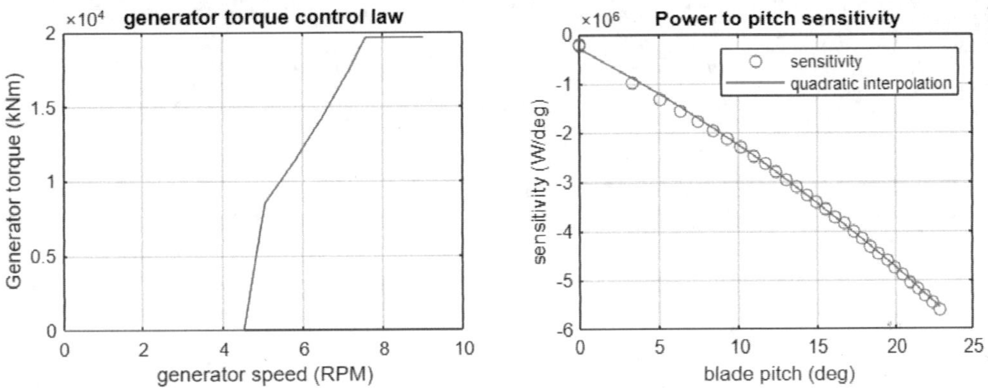

***Figures 4**: generator torque control law implemented for the IEA 15 MW wind turbine and the power to pitch sensitivity and the quadratic interpolation used in the gain scheduled PI control system*

The blade pitch control is used for the power regulation above the rated wind speed and uses a gain scheduled PI control. The equation implemented for the blade pitch control is described by the following equations:

$$\theta = K_p \Delta\omega + K_I \frac{\Delta\omega}{s} \quad (1)$$

$$\Delta\omega = \begin{cases} \omega - \omega_0 & if\ \omega > \omega_0 \\ 0 & if\ \omega \leq \omega_0 \end{cases} \quad (2)$$

Where θ is the blade pitch, ω is the generator speed, ω_0 is the rated generator speed (7.56 RPM), K_P and K_I are the gains for the PI control.

$$K_P = \frac{K_P'}{\frac{dP}{d\theta}} \quad (3)$$

$$K_I = \frac{K_I'}{\frac{dP}{d\theta}} \quad (4)$$

$$\frac{dP}{d\theta} \cong c_1\theta^2 + c_2\theta + c_3 \quad (5)$$

$\frac{dP}{d\theta}$ is the power to pitch sensitivity and c_1, c_2 and c_3 are the coefficients for the parabolic equation that interpolates the power to pitch sensitivity. The values of K_P' and K_I' are obtained from an optimisation algorithm using MOST for the minimisation of the pitch rate of the platform. The same controller and parameters have been implemented in FAST.

Control parameters	c_1	c_2	c_3	K_p'	K_I'
values	2500 W/deg^3	173436 W/deg^2	290832 W/deg	49384000 s	3938500

Table 2: *parameters for the baseline control implemented on IEA 15 MW wind turbine*

3. Results

To make a comparison between Fast and Most, we have considered a Volturn US platform [14], which supports the IEA 15 MW wind turbine [15]. Moordyn is used to simulate the mooring forces. The degrees of freedom of MOST and FAST are surge, heave and pitch. The wave forces are not considered (still water condition). The hydrodynamic properties of the platform are computed using Nemoh [16] and then imported to WEC-Sim as .h5 file. The time step of simulation is 0.05 seconds. Since the boundary element method does not consider the viscous forces, the quadratic drag of the structure needs to be estimated with an external software. In FAST, the quadratic drag matrix for the six degrees of freedom is estimated using OpenFOAM. Such matrix is then imported in WEC-Sim.

In the figures (5), the pitch of the platform and the power production are shown for two different mean wind speeds, under and above the rated wind speed (10.59 m/s): 8 and 14 m/s. additionally, in the figures (6) the axial force and moment are shown. The blade root forces consider all the forces acting on blade, including aerodynamics, gravitational forces, inertial forces, and gyroscopic moments. The main oscillations in the blade root forces are caused by the wind shear, the turbulence on the wind, and the periodic oscillations of the gravity moment due to the precone and tilt angle.

In this paragraph, a comparison of the time simulation is included (table 3). The choice of the mooring model is determinant in terms of computation time, since the choice of Moordyn takes about three times longer respect to the quasistatic matrix. The time of initialisation and execution of the Simscape model is three-five times longer than FAST, but it could be further reduced by generating a C code starting from the Simscape model.

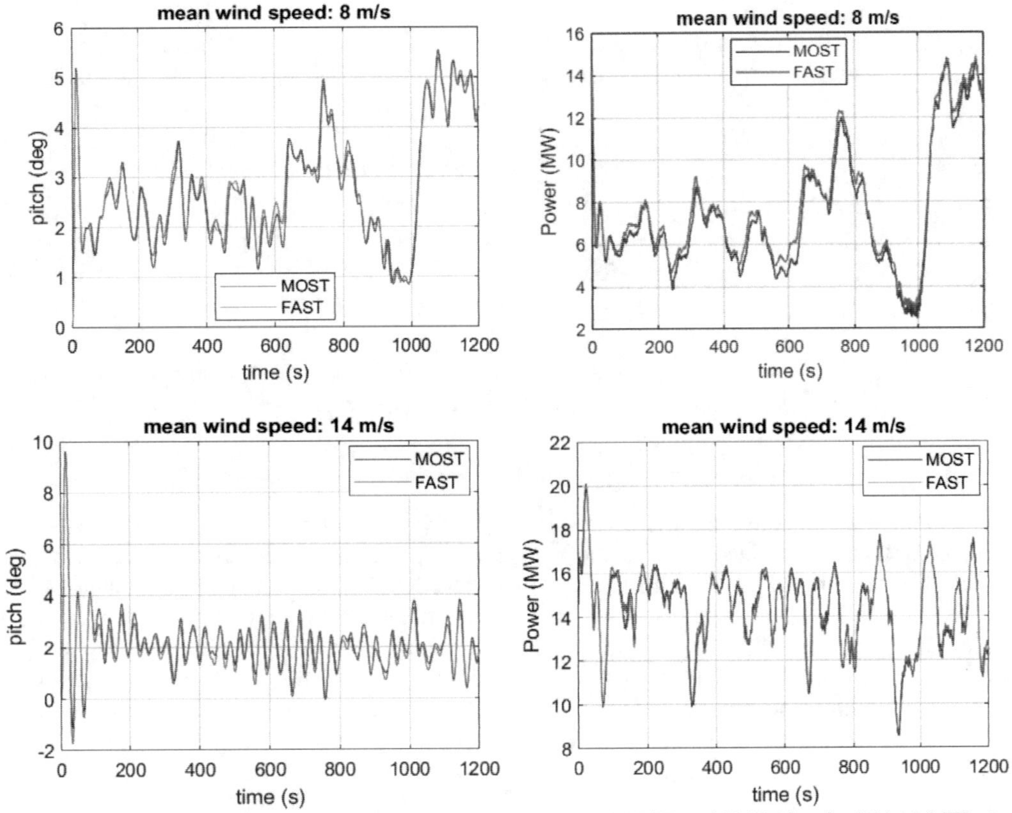

***Figures 5**: Platform pitch and power output comparison between MOST and FAST for the IEA 15 MW wind Turbine*

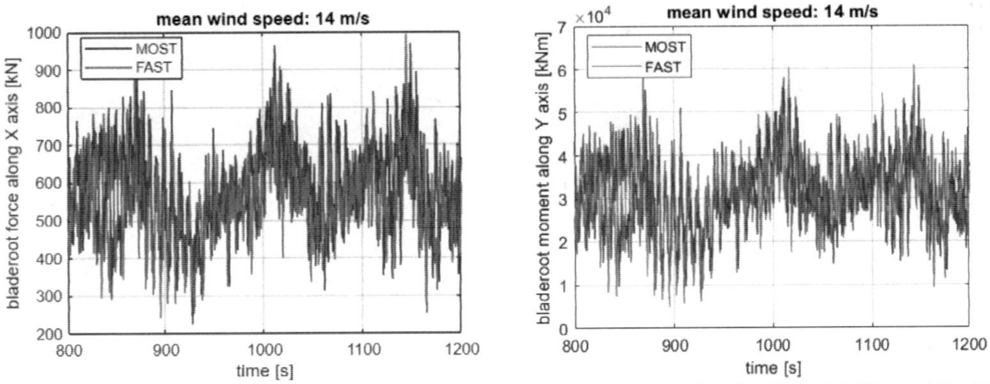

***Figure 6**: axial force and moment comparison between MOST and FAST for the IEA 15 MW wind Turbine*

Simulation time (seconds)	Computation time (seconds)		
	FAST	MOST (Moordyn)	MOST (quasi-static mooring matrix)
1 (initialisation)	11	25	21
1200	122	106	40
3600	366	264	72

***Table 3**: Simulation time comparison between FAST and MOST. The two options for the mooring have a significant impact on the simulation time. The time step of simulation is 0.05 seconds.*

Mean wind speed (m/s)	Pitch (deg)				Power (MW)			
	Mean		Standard deviation		mean		Standard deviation	
	MOST	FAST	MOST	FAST	MOST	FAST	MOST	FAST
5	0.527	0.492	0.370	0.355	1.756	1.860	0.921	0.866
8	2.738	2.794	1.142	1.129	7.383	7.726	2.905	2.876
11	3.469	3.263	1.143	1.111	13.63	13.83	2.421	2.245
14	2.112	1.942	1.060	1.046	14.38	14.40	1.848	1.798
17	1.522	1.369	1.127	1.167	14.42	14.43	1.910	1.891
21	1.141	0.996	1.341	1.336	14.58	14.63	2.492	2.449

***Table 4**: Platform pitch and power output comparison between MOST and FAST for different mean wind speeds. The mean value and standard deviation are used as comparison objects.*

4. Conclusion

The calculation of the aerodynamic forces carried out by the Aerodyn module represents the greatest cost in computational terms, because the pressure on the blade is computed at each node and it involves an iteration cycle. The simplification of aerodynamics through look-up tables using pre-calculated values is a big advantage for reducing the simulation time. In fact, the simulation time of the Simscape model is about three-five times faster than FAST. The results of the model however show a good agreement compared to the FAST results, considering the blades and tower as rigid bodies and neglecting other minor effects, like the tower shadowing and the wake dynamics. These simplifications are necessary to reduce the time computation, which is the primary goal of the model. Indeed, this model is intended to be used in optimisation algorithms for the platform design, control system parameters and wave energy converters in hybrid platforms, where the flexibility, expansibility, and computation cost are crucial, being able to explore preliminary designs in an efficient way, that are not possible with other codes like FAST. The model however leaves out important dynamic effects, especially the flexibility of the blades, that can be important for the load cycles. Such issue can be addressed in new developments of the model, for example introducing the modal analysis for the blades and the tower, modified and adapted to the usage of the look-up tables for the aerodynamic forces.

References

[1] Joao Cruz and Mairead Atcheson, *Floating Offshore Wind Energy*. Cham: Springer International Publishing, 2016. doi: 10.1007/978-3-319-29398-1.

[2] J. M. Jonkman, "Dynamics Modeling and Loads Analysis of an Offshore Floating Wind Turbine," 2001.

[3] A. J. Coulling, A. J. Goupee, A. N. Robertson, J. M. Jonkman, and H. J. Dagher, "Validation of a FAST semi-submersible floating wind turbine numerical model with DeepCwind test data," *Journal of Renewable and Sustainable Energy*, vol. 5, no. 2, p. 023116, Mar. 2013, doi: 10.1063/1.4796197.

[4] F. Driscoll, J. Jonkman, A. Robertson, S. Sirnivas, B. Skaare, and F. G. Nielsen, "Validation of a FAST Model of the Statoil-hywind Demo Floating Wind Turbine," *Energy Procedia*, vol. 94, pp. 3–19, Sep. 2016, doi: 10.1016/J.EGYPRO.2016.09.181.

[5] J. R. Browning, J. Jonkman, A. Robertson, and A. J. Goupee, "Calibration and validation of a spar-type floating offshore wind turbine model using the FAST dynamic simulation tool," *Journal of Physics: Conference Series*, vol. 555, p. 012015, Dec. 2014, doi: 10.1088/1742-6596/555/1/012015.

[6] J. Kim and H. Shin, “Validation of a 750 kW semi-submersible floating offshore wind turbine numerical model with model test data, part II: Model-II,” *International Journal of Naval Architecture and Ocean Engineering*, vol. 12, pp. 213–225, Jan. 2020, doi: 10.1016/J.IJNAOE.2019.07.004.

[7] R. Damiani and M. Franchi, “An innovative second-order design method for the structural optimization of the SpiderFLOAT offshore wind Platform,” *Ocean Engineering*, vol. 228, p. 108792, May 2021, doi: 10.1016/J.OCEANENG.2021.108792.

[8] Ogden N, Ruehl K, Yu Y H, and A. Keester, “Review of WEC-Sim Development and Applications.,” *Proceedings of the 14th European Wave and Tidal Energy Conference, EWTEC 2021, Plymouth, UK*, 2021.

[9] Hansen M H, Hansen Anca Daniela, Larsen Torben J., Øye S, Sørensen P, and Fuglsang P, “Control design for a pitch-regulated, variable speed wind turbine,” *Forskningscenter Risoe*, 2005.

[10] N. D. Kelley, “Turbulence-Turbine Interaction: The Basis for the Development of the TurbSim Stochastic Simulator,” *National Renewable Energy Laboratory*, 2011.

[11] K. Ruehl, C. Michelen, B. Bosma, and Y.-H. Yu, “WEC-Sim Phase 1 Validation Testing: Numerical Modeling of Experiments,” 2016.

[12] M. Lawson, B. B. Garzon, F. Wendt, Y.-H. Yu, and C. Michelen, “COER Hydrodynamic Modeling Competition: Modeling the Dynamic Response of a Floating Body Using the WEC-Sim and FAST Simulation Tools,” 2015.

[13] Masciola Marco, Jonkman Jason, and Robertson Amy, “Implementation of a Multisegmented, Quasi-Static Cable Model,” *Twenty-third International Offshore and Polar Engineering Conference, Anchorage, Alaska*, 2013.

[14] C. Allen *et al.*, “Definition of the UMaine VolturnUS-S Reference Platform Developed for the IEA Wind 15-Megawatt Offshore Reference Wind Turbine Technical Report,” 2020.

[15] E. Gaertner *et al.*, “Definition of the IEA Wind 15-Megawatt Offshore Reference Wind Turbine Technical Report,” 2020.

[16] Babarit Aurélien and Delhommeau Gérard, “Theoretical and numerical aspects of the open source BEM solver NEMOH,” 2015.

Journal of Physics: Conference Series 2257 (2022) 012004 doi:10.1088/1742-6596/2257/1/012004

Digitalization Workflow for Automated Structuring and Standardization of Maintenance Information of Wind Turbines into Domain Standard as a Basis for Reliability KPI Calculation

Marc-Alexander Lutz[1], **Julia Walgern**[2], **Katharina Beckh**[3], **Juliane Schneider**[2], **Stefan Faulstich**[1], **Sebastian Pfaffel**[1]

[1]) Fraunhofer Institute for Energy Economics and Energy System Technology IEE
Königstor 59, 34119 Kassel, Germany
[2]) Fraunhofer Institute for Wind Energy Systems IWES
Postkamp 12, 30159 Hanover, Germany
[3]) Fraunhofer Institute for Intelligent Analysis and Information Systems IAIS
Schloss Birlinghoven, 53757 Sankt Augustin, Germany

E-mail: alexander.lutz@iee.fraunhofer.de

Abstract. Maintenance data of wind turbines is an important information source for calculating key performance indicators. Also, it can be used for developing models for early fault detection. Both activities aim for supporting informed decisions in operation and maintenance. However, such data is rarely available in a structured and standardized format which hinders the interoperability of different enterprises. Consequently, maintenance information is often unused or only usable with considerable personnel effort. To digitalize wind farm maintenance, a digitalization workflow is developed and presented in this paper. The workflow consists of the steps optical character recognition, information extraction and text classification. The workflow is applied on real-world wind turbine service reports and invoices. First results for each step show good performance metrics and potential for further real-world application of the proposed method.

1. Introduction

Up to 30% of the *Levelized Cost of Energy* (LCOE) of *Wind Turbines* (WTs) are driven by the cost of *Operation and Maintenance* (O&M) [1]. In order to further decrease this share and to be able to perform informed and strategic decisions in the maintenance of WTs, Key Performance Indicators (KPIs) based on maintenance information are crucial. Modern enterprises are embedded in a fast-paced environment which is dependent on data. In order to use this data and to ensure interoperability of different enterprises, a common structure and standards are necessary. Even though there are already standards and guidelines available for the communication of maintenance data and to classify affected components of a WT during a maintenance measure, such as Reference Designation System for Power Plants® (RDS-PP®) [2] or Reference Designation System for Power Systems (RDS-PS) [3], they have only been applied to a limited extent to date. This results in a variety of maintenance reports and formats and hinders communication between enterprises. Consequently, maintenance information is heterogeneous,

unstructured and rarely standardized. Since utilization of maintenance information is only possible with considerable personnel effort, KPI-driven maintenance optimization or developing models for early fault detection is hindered and scarce within the wind energy domain.
Even if standards are implemented at some point, historical data can not be used for further analysis or reliability KPI calculation yet.
To be able to use historical data and to speed up the usage of standards in the maintenance of WTs, a workflow for digitalization of maintenance information (digitalization workflow) (see Figure 1) is proposed. The digitalization workflow consists of the three steps optical character recognition, information extraction and text classification. To validate the workflow, real-world WT service reports and invoices are used.
The outline of the paper is as follows: After the introduction (Section 1) the state of the art on digitalization methods including optical character recognition, information extraction and classification is presented in Section 2. Furthermore, an overview of the challenges with current maintenance data of WTs and of available standards is given. In Section 3 the methods used for comparing WT service reports and for implementing the digitalization workflow are outlined. Also, the data set utilized for validating the digitalization workflow is described. Afterwards, the results for the aforementioned methods are presented and briefly discussed (Section 4). In Section 5 the findings of the paper are summarized and an outlook is given.

2. State of the Art on Digitalization Methods and Existing Standards

After describing the current situation on how maintenance information is recorded and stored within the wind industry, different approaches for digitalizing data are reviewed. Afterwards, the following sub-chapters revise the state of the art of the individual steps of the proposed digitalization workflow.

2.1. State of the Art on Recording Maintenance Information

Maintenance information are available in numerous maintenance documents. File formats can vary from documents which are structured already (Excel, CSV) to scans of letters in pdf or image format. Dependent on the document type, the depth of information about the maintenance measure differs: While maintenance reports or turbine logs can provide detailed failure and measure descriptions, invoices often give limited technical insights. A large heterogeneity of documents complicates the data processing and not from all documents the same information content can be extracted. Moreover, even within one type of document the depth of information can vary significantly. This is mainly related to data sovereignty: Dependent on which *Service Enterprise* (SE) is performing the maintenance service at the WT and which documentation has been agreed on, the information content differs substantially. A reduced information content hampers labeling the maintenance measure and the related WT component. Standardization of different document types and their content is crucial to evaluate and communicate the data in an automated manner. As the majority of maintenance information is described in free text, text classification is required to label the affected component consistently and to make the information comparable (cf. [4]).

2.2. State of Research on Approaches for Digitalization of Maintenance Information

Recommendations on how to collect data for the purpose of reliability assessment are summarized in [5]. However real-world data is often not collected according to this recommendation. Therefore, solutions to make use of this data are necessary. Approaches that aim to extract knowledge from natural language in maintenance data of different domains are summarized under the term Technical Language Processing [6]. Discovering KPIs from natural language in maintenance work orders is shown in [7]. Gao et al. focus on the extraction of performance

metrics, e.g. mean time to failure, and therefore implement a pipeline for machine reading of unstructured maintenance work orders [8].
Approaches that are specific to the wind energy domain can be seen in [9, 10, 11]. Blanco-M. et al. show a text mining approach to asses the failure condition of WTs by using the maintenance history [9]. A domain ontology for wind energy is presented in [12]. Other methods combine WT sensor data and maintenance reports for the purpose of predictive maintenance [11]. Further approaches focus on textual information of accidents at WTs to generate knowledge [13, 14]. However, none of the before mentioned approaches consider domain standards like RDS-PP® [2] or RDS-PS [3] for describing the affected components in a uniform manner.

2.3. *Optical Character Recognition*
Optical Character Recognition (OCR) is a well researched field for more than six decades [15, 16]. Several reviews and books within this domain are available [16, 17]. An evaluation of the performance metrics of various OCR approaches is provided in [18]. Different OCR tools such as ABBYY FineReader [19], OmniPage [20] or Tesseract [21] exist. A comparison of Tesseract to other tools is shown in [22].

2.4. *Information Extraction*
Information extraction aims at creating a structured representation of selected information out of text [23]. Information extraction is a major task in computer linguistics that is subdivided in several areas, e.g. template filling, named entity recognition, identification of relations and event detection [24, 25]. An overview of the domain-specific information extraction is given in [23, 26]. The various applications are shown in [27].
Approaches to extract content and to derive relations out of PDF documents are shown in [28, 29].

2.5. *Classification*
Classification in the machine learning sense is the task to automatically assign categories to data points. Depending on the task, it can be distinguished between assigning one class to a data point (multi-class) and assigning several classes (multi-label). The domain standard that is considered in this work, RDS-PP®, contains a number of categories up to the four-digit range. Thus, relevant classification methods need to be able to tackle this amount of categories. Two types of classification models are suitable. First, Support Vector Machine (SVM) is an one-vs-all approach that uses one classifier for each class [30, 31]. Originally intended to perform binary classification (assigning one out of two classes) the SVM can be employed for multi-label as well. Second, word embedding approaches map high dimensional and sparse co-occurrence matrices based on the words that occur in the training data to a low dimensional latent vector space [32]. Most current approaches make use of this method. Language models such as BERT and other transformer variants [33, 34, 35] use pre-trained models and fine-tune on given data. They can be used, among other applications, to perform classification tasks.

3. Methodology
In this section the methodology for each step within the digitalization workflow and the data sets used for validation are described. First, it is explained how wind turbine service reports are compared (Subsection 3.1). Second, the data sets used for validating the different steps of the proposed digitalization workflow (Subsection 3.2) are described. Third, an overview of the implementation of the digitalization workflow is given (Subsection 3.3).

3.1. Comparison of existing Field Data: Wind Turbine Service Reports

An overview of initiatives that collected maintenance data of WTs for the purpose of reliability studies is shown in [36, 37, 38]. Acquisition of data but especially evaluation is most efficient if same formats and structures are used. This can be the case for maintenance data being collected and structured by initiatives or for maintenance data recorded by the same SEs. As stated before, in real world a variety of different templates is used. In order to understand which information content is available and needs to be considered within the digitalization workflow, several service reports of WTs are compared and the available information content provided by different SEs is clustered into categories such as site name, WT-type or the text description of the maintenance measure (TDoMM).
The data set for this comparison consists of five different types of service reports which have been recorded by five different SEs.

3.2. Data Sets used for the Validation of the Digitalization Workflow

Two different data sets are used to evaluate the steps of the digitalization workflow.
The first data set "invoices" is used for the validation of the steps OCR and information extraction. It consists of 152 invoices that have been issued by one SE to an operator of WTs. The invoices are available as scanned images. For the purpose of enterprise resource planning (ERP) the relevant information of the invoices, such as the issued date, the invoice number and the invoice total, are currently transferred manually into the ERP system. Next to the invoices, this information is also available. Therefore, the manually inserted information from the ERP system can be used as a ground truth to evaluate the performance of the first two steps of the digitalization workflow.

The second data set "service reports" is used for the steps of information extraction and classification. It comprises of 4000 offshore WT service reports covering 240 operational turbine years. The service reports are provided in PDF format and contain the date of the measure, the number of technicians, the working hours, the materials used as well as the TDoMM. Initially, the TDoMM is not labeled with a domain standard such as RDS-PP®. As a first step, the labeling according to RDS-PP® is performed manually by a domain expert. Each TDoMM is labeled with one or more RDS-PP® component categories depending on how many measures are undertaken at different WT components. Once all available TDoMMs within the WT service reports are labeled according to RDS-PP®, a basis for training and testing the classifier is available.

3.3. Digitalization Workflow

The digitalization workflow, depicted in Figure 1, consists of the three steps: OCR, information extraction and text classification. The different steps of the digitalization workflow are theoretically described in Lutz et al. [4] and are briefly outlined again.

At different stages of the digitalization workflow, structured information can be archived. This can be the case following the steps of OCR and information extraction or after the classification step, which even allows for structured and standardized information being stored. Since archiving can be easily automated, this step is illustrated in Figure 1 as part of the digitalization workflow but is not explained in the following.
If maintenance information is present in the form of pictures or scans, at first it is necessary to apply OCR. By doing so, scans are converted, the output thereafter is machine-readable text. Next to providing semi-structured files like .csv or .txt, the output of the OCR can serve as input for the step of information extraction. The purpose of information extraction is to convert multiple different files into one common structure where in same columns same information

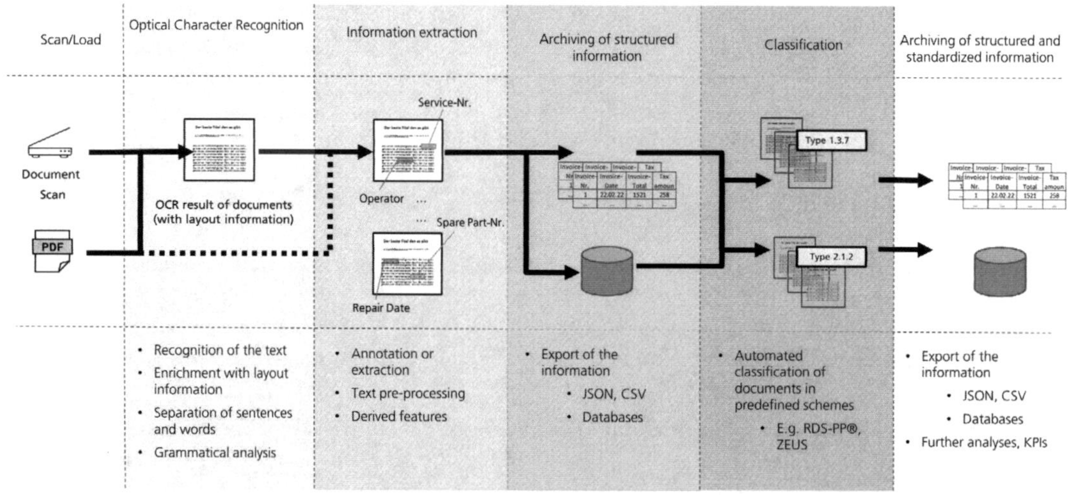

Figure 1: Overview of the proposed digitalization workflow

categories can be found, e.g. the date of the maintenance measure or the TDoMM. Thus, many different service reports can be converted into a single tabular form, e.g. a spread sheet. With the aforementioned steps, structured information is available. Since some information categories, such as the TDoMM, are often given as free-text information, those information still need to be standardized utilizing predefined categories. By using text classification methods, this text can be classified into categories which are given by a domain standard (e.g. RDS-PP® [2], RDS-PS [3] or State-Event-Cause-Code (ZEUS) [39]). This results in structured and standardized maintenance information being available and usable for further analyses.

3.3.1. Optical Character Recognition

Maintenance data can be available in different formats. If data is available as scanned images or pictures, information can neither be extracted nor can the content of the data be searched. To be able to access the information, OCR is applied which converts images into text. The text can then be searched and further used. The OCR output can be used as input for the next digitalization workflow step. As stated in Section 2, OCR is a well researched domain. Therefore, OCR can easily be implemented into the digitalization workflow. In this work the authors use the OCR open source engine Tesseract [21] as it provides a good performance in comparison to other tools [22].

3.3.2. Information Extraction

Besides the output of the OCR, semi-structured files like .csv, .pdf or .txt can be used as input for the step of information extraction. In comparison to the definition of information extraction within the field of computer linguistics (see Section 2), information extraction within the proposed digitalization workflow has the purpose of converting multiple different single files into one common structure in which same information categories can be found in same columns, e.g. the date of the maintenance measure or the TDoMM.

Information extraction within the proposed workflow comprises of the sub-steps template filling and text pre-processing. The sub-step of text pre-processing is similar as outlined in [40]. Within the sub-step of template filling each part of the text is assigned to an information category which allows for transferring information in the same structure. With the aforementioned sub-steps, structured information is available. However, some information categories still need to be standardized, for instance the TDoMM. This can be achieved by using text classification methods, which are described in the next Section.

3.3.3. Classification

The data basis used for the classification is the data set "service reports" containing around 4000 service reports of WTs which are manually labeled according to RDS-PP® by a domain expert. RDS-PP® is hierarchically structured (see Appendix, Figure 4). An example for a possible classification of the TDoMM in WT service reports to RDS-PP® is given in Table 3 in the Appendix. By analyzing the data set it is found that on RDS-PP® level two, around 55% of all service reports are represented by only three RDS-PP® labels (see Figure 2). Therefore, the data set is imbalanced.

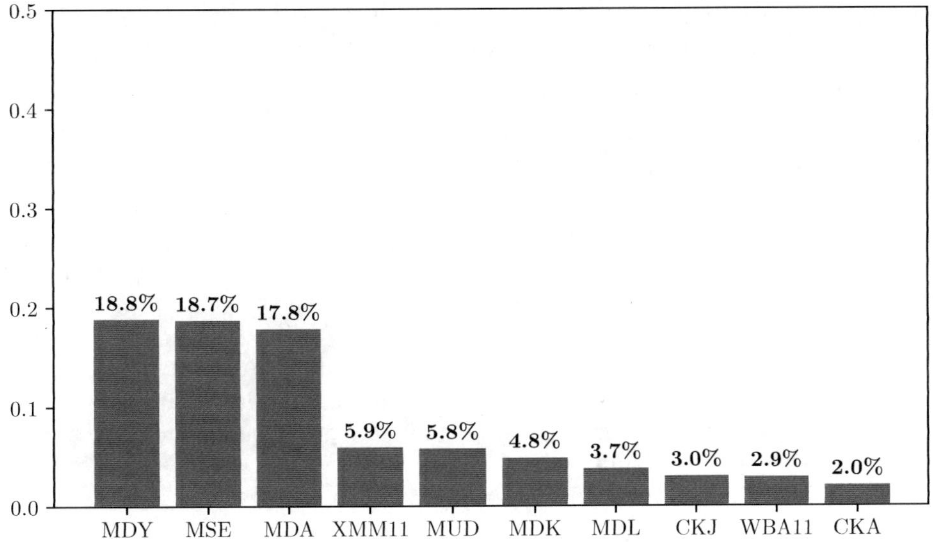

Figure 2: Distribution of the ten most frequent RDS-PP® labels on level 2 within the data set "service reports"

While on RDS-PP® level one, 16 different categories are theoretically available, on level two 62 RDS-PP® categories can be used. Therefore, typical classifier choices are established methods such as SVM [30] and transformer variants such as BERT [33]. Transformer variants require a large corpus of domain-specific text which is not available for WTs. Consequently, SVM is used as classification method. Within the data set each TDoMM is labeled with one or more RDS-PP® components, which makes it a multi-label classification.
In general, the text classification consists of four steps:

- **Text cleaning**: E.g. removing special characters and lowercasing
- **Vectorization**: Each report is represented by a vector

- **Training**: Splitting the data in train and test set and training a classifier on the train set to learn a mapping from text to RDS-PP® label(s)
- **Evaluation**: Test the classifier on the remaining test set

Each TDoMM is transformed with TF-IDF (term frequency-inverse document frequency) [40] which assigns a value of information content to each word. This results in low values for frequent words, e.g. "the", and high values for relatively infrequent words, e.g. "nacelle".
In particular, for each RDS-PP® level one SVM is trained such that performance metrics for each level can be compared. In addition, one classifier is trained which is referred to as *final*. For the final classifier all given labels on all respective levels are used for training and prediction. This classifier allows for labeling all TDoMMs without knowing which RDS-PP® level is searched for and primarily serves as a reference point for now.
Evaluation is performed with F1-scores. F1 is calculated from recall and precision as follows

$$F_1 = 2 \times \frac{\text{precision} \times \text{recall}}{\text{precision} + \text{recall}} = \frac{2 \times \text{TP}}{2 \times \text{TP} + \text{FP} + \text{FN}}$$

Herein, TP, FP and FN refer to true positives, false positives and false negatives, respectively. F1 is a metric that is more robust towards class imbalance. Thus, it is the preferred choice over other metrics such as accuracy. A distinction can be made between micro F1 which takes label frequencies into account and macro F1 that treats all labels equally.

4. Results and Discussion

4.1. Comparison of existing Field Data: Wind Turbine Service-Reports

Different service reports of five SEs are compared. Results can be seen in Table 1. It can be noticed that some information categories are available in all service reports, e.g. the type and the identifier of the WT, the date and the TDoMM. However, the description of the TDoMM is not documented in a standardized way, for instance by using RDS-PP® [2], RDS-PS [3] or ZEUS [39]. Instead, free text is used. Different information categories can be seen in service reports of different SEs. Also the type of documentation format varies. The service reports are available as scanned images, Word documents, XML or PDF-files.
The results of this comparison (see Table 1) support the assumption that maintenance data provided by different SEs differ significantly regarding the structure and information content. Furthermore, standards for classifying maintenance activities are rarely used. This raises the challenge as well as the demand of an automated digitalization workflow.

Table 1: Available information categories in service reports of different SEs

	Site Name	Costumer	WT Type	WT Number	Service Number	Energy Produced	Operational Time	Maintenance Reason	Wind Speed at arrival	TDoMM	Material Consumption	Time of Maintenance	Open Tasks	Comments	Technicians	Working Hours	Date Next Measure	Status of Work Order	Commuting Distance	ID Work Vehicle	Contract Number
SE 1	X	X	X	X	X	X	X			X	X	X		X	X	X	X	X			
SE 2	X		X	X	X				X	X			X	X	X						
SE 3	X	X	X	X	X			X		X	X	X			X	X					
SE 4	X	X	X	X	X			X		X	X	X			X	X					
SE 5	X	X	X	X	X	X		X		X	X	X				X			X	X	X

4.2. Digitalization Workflow

In the following sections results for the different steps of the digitalization workflow are presented.

4.2.1. Optical Character Recognition

OCR is conducted on the first data set "invoices". As described in Section 3.2, 152 invoices are available as scanned images. After using OCR all invoices are converted to machine-readable text. Since an automated evaluation of the feasibility of this process is only possible after the step of information extraction, results are described in the following Section 4.2.2.

4.2.2. Information Extraction

After the step of OCR the data set "invoices" is further processed using the step of information extraction. As a result, all invoices are stored in the same structured format. This allows for comparing and validating the extracted information categories with the ground truth data exported from the ERP-system (see Section 3.2). The results can be seen in Table 2.

Table 2: Results of OCR and information extraction for the first data set "invoices"

Number of invoices evaluated	152
Share of correct entries [%]	94.5

Next to the number of invoices evaluated, the share of correct entries is shown. One entry is defined as the content of a certain information category in one invoice such as the issuing date of the invoice or the amount of cost stated in the invoice. Out of all available entries being extracted a share of 94.5% is similar as listed in the ground truth and is therefore evaluated as "correct".

Furthermore, information extraction is performed on the data set "service reports". All different information categories in the WT service reports of the SE could be extracted and summarized in a common table. The content of each report is represented in a row of that common table.

Same information categories are found in same columns. No ground truth data is available to validate automatically if all data is extracted correctly. Nevertheless, after carefully comparing the content of many service reports with the structured result manually, the authors believe that all service reports are extracted correctly.

4.2.3. Classification

For the TDoMMs of the data set "service reports" single-label and multi-label classifiers for different RDS-PP® levels are trained and tested. Classification results are depicted in Figure 3. Each level represents a level in the RDS-PP® hierarchy. Figure 3(a) and (b) show results for multi-label and Figure 3(c) and (d) for single-label classification. Figure 3(a) and (c) refer to micro F1 while Figure 3(b) and (d) depict macro F1-scores. In both cases, multi-label and single-label, macro F1-scores are lower than micro F1-scores. This can be explained by the label imbalance in the data set, i.e. that some labels occur more frequently than others (also see

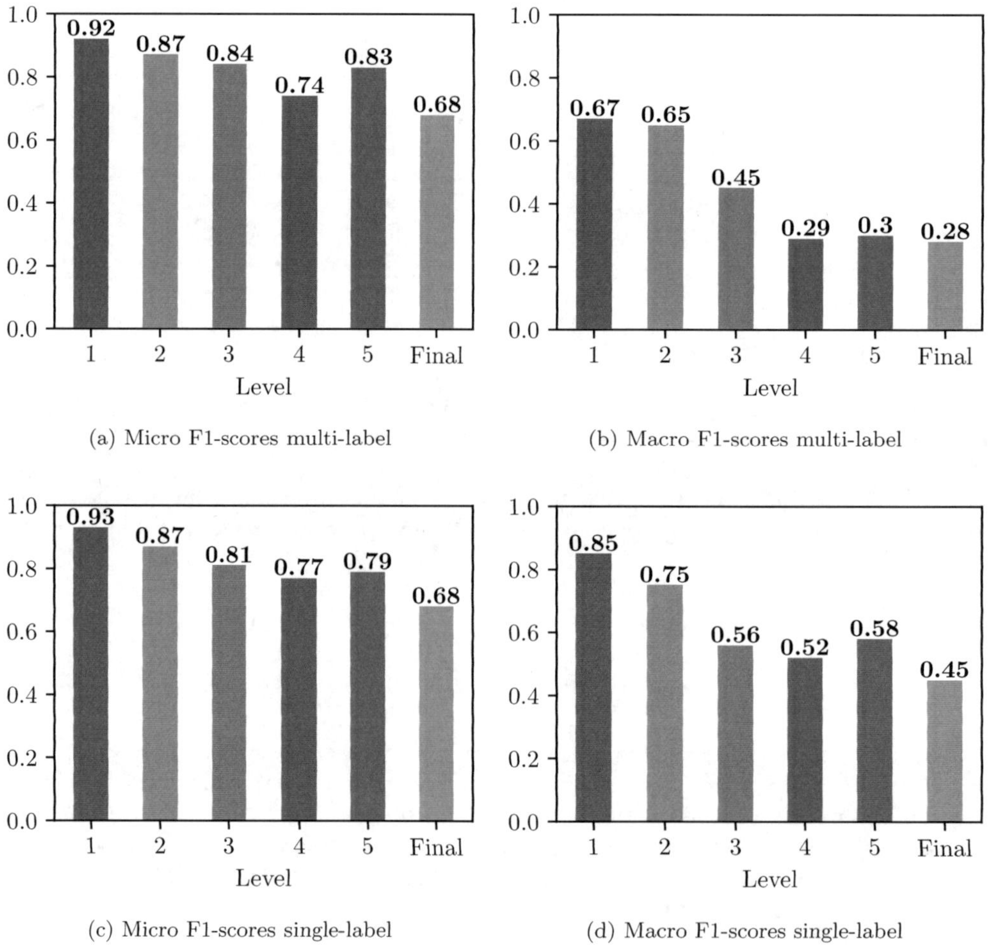

Figure 3: Micro and macro F1-scores of single-label and multi-label classifiers for different RDS-PP® levels

Figure 2). As a result, infrequent labels cannot be captured well by the classifier and the micro F1-scores are higher than the macro F1-scores. In the multi-label as well as in the single-label setting it can be seen that with ascending levels (from one to five) F1-scores decrease. This is expected since the number of data points that belong to a certain level decrease from level one to level five, while, at the same time, the level of annotation detail is increasing. Only classifier results for level five show outliers in this descending trend. Even though more RDS-PP® labels are theoretically available for labeling the TDoMM in WT service reports, only 16 different labels are used for level five within the data set. In comparison, over 70 labels are used for labeling on level four as information depth is rarely precise enough to label until level five. Therefore, the actual label space in level five is smaller than e.g. for level four leading to better classifier performance for level five. Inspecting the macro F1-scores, the results show higher scores for the single-label than for the multi-label setting. This can be explained by the fact that the single-label setting reduces the problem to predicting only one affected WT component because multi-label cases are not included in the training and test set. This effect cannot be observed when comparing the micro F1-scores which needs to be analyzed further.
Overall, the results indicate that a classification system can provide decision support in the annotation process and for certain RDS-PP® levels an automatic labeling procedure is possible.

5. Conclusion & Outlook

WT service reports are often heterogeneous. Their information content and information depth vary. Also, no uniform structure or standards are used for documentation. Data preprocessing as well as determining the maintained component are therefore essential steps for enabling further analyses. At the same time, these steps can only be performed manually so far, thus are time-consuming, and often require the knowledge of a domain expert.
The present paper proposes a digitalization workflow for maintenance information of WTs including the steps of OCR, information extraction and classifcation. Since OCR and information extraction show good results with an accuracy of around 95% or above depending on the data set evaluated, the focus lies on classification of maintained components according to the standard RDS-PP®. The process uses a SVM for text classification on a real world data set. First classification results look promising with micro F1-scores showing high values. However, as not every component is maintained in the same frequency, the RDS-PP® label distribution in the data set is imbalanced. Thus, higher represented labels are captured better by the classifier. A lower macro F1-score is therefore seen. Beside the label distribution, investigations show that the level depth of assigned RDS-PP® labels also has an impact on the classification performance. The digitalization workflow allows utilizing operational knowledge which was not accessible in the past. Furthermore, the results indicate that automated digitalization of maintenance reports of WTs is possible and the usage of standards can be accelerated. Thereafter, KPIs can be calculated automatically which supports data-driven decision making.
Further analyses and uncertainty evaluations for different data sets are planned to drive decisions for productive use. Also, classification for other guidelines and standards, e.g. ZEUS or RDS-PS, will be tested. The comparison of different classifiers and associated insights will be utilized to develop classifiers further considering application-specific requirements.

6. Acknowledgements

This research is to a large part carried out within the research project "Digitalization of Maintenance Information" (DigMa) funded by the German Federal Ministry of Economic Affairs and Climate Action (BMWK), grant number 03EE2016A. The authors thank the project partners for providing field data and for sharing their requirements and experience.

7. Appendix

7.1. Structure of RDS-PP®

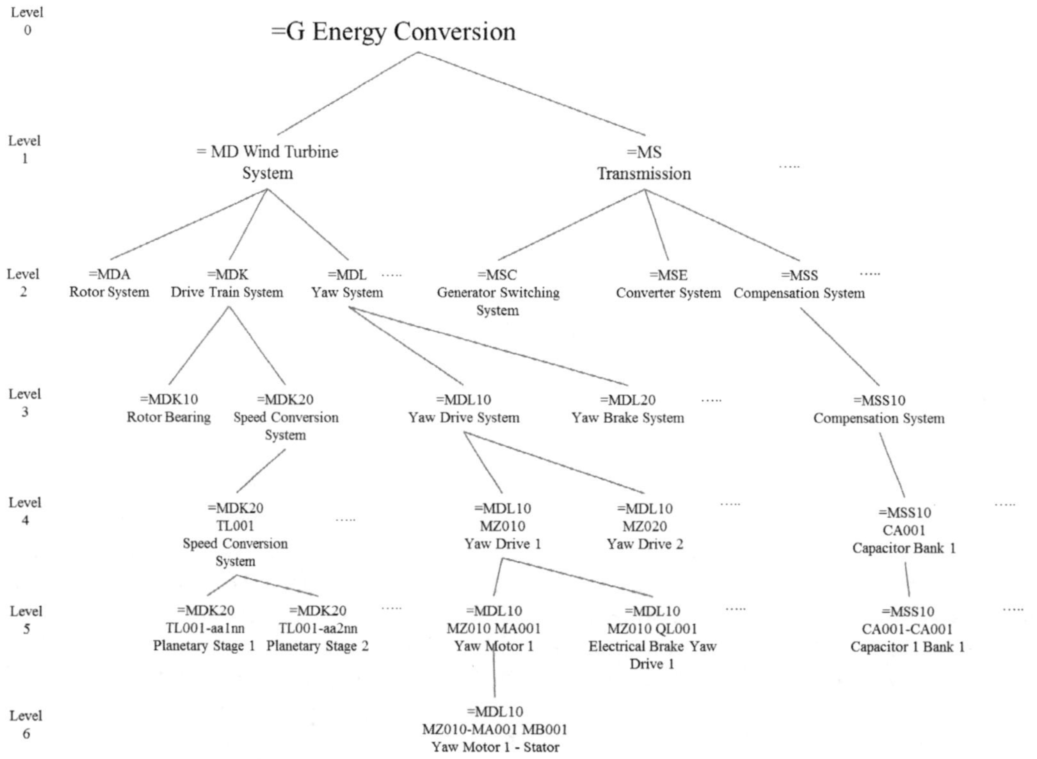

Figure 4: Overview of the hierarchical structure of RDS-PP® [2, 4]

7.2. Exemplary classification

An example for a possible classification of the TDoMM in WT service reports to RDS-PP® is given in Table 3.

Table 3: Example for the classification of the TDoMM according to RDS-PP®

	Measure 1	Measure 2
TDoMM	Inspection of surface damage at azimuth system.	Failure due to icing at the rotor blade one. WT shut down.
RDS-PP® Label	Yaw System	Rotor Blade System 1
RDS-PP® ID	=MDL	=MDA11

References

[1] International Renewable Energy Agency, "Renewable power generation costs in 2019." `https://www.irena.org/publications/2020/Jun/Renewable-Power-Costs-in-2019`, 2020.

[2] VGB PowerTech e.V., "Reference Designation System for Power Plants® - Application Guideline; Part 32: Wind Power Plants." Book, 2014.

[3] Danish Standards Foundation, "A Guide to RDS - Reference Designation Systems. TAG Numbers for Systems in Accordance with the ISO/IEC 81346 Standard Series," 2020.

[4] M. A. Lutz, K. Beckh, J. Kindermann, J. Schneider, J. Walgern, S. Pfaffel, S. Faulstich, and A. Staack, "Digitalisierungsworkflow zur Strukturierung und Standardisierung von Instandhaltungsinformationen von Windenergieanlagen," *Gesellschaft für Informatik; Lecture Notes Informatics*, pp. 229–249, 2021.

[5] B. Hahn, T. Welte, S. Faulstich, P. Bangalore, C. Boussion, K. Harrison, E. Miguelanez-Martin, F. O'Connor, L. Pettersson, C. Soraghan, C. Stock-Williams, J. Dalsgaard Sørensen, G. van Bussel, and J. Vatn, "Recommended practices for wind farm data collection and reliability assessment for o&m optimization," *Energy Procedia*, vol. 137, pp. 358–365, 2017.

[6] M. P. Brundage, T. Sexton, M. Hodkiewicz, A. Dima, and S. Lukens, "Technical language processing: Unlocking maintenance knowledge," *Manufacturing Letters*, vol. 27, pp. 42–46, 2021.

[7] M. Navinchandran, M. E. Sharp, M. P. Brundage, and T. B. Sexton, "Discovering critical KPI factors from natural language in maintenance work orders," *Journal of Intelligent Manufacturing*, 2021.

[8] Y. Gao, C. Woods, W. Liu, T. French, and M. Hodkiewicz, "Pipeline for Machine Reading of Unstructured Maintenance Work Order Records," in *Proceedings of the 30th European Safety and Reliability Conference and 15th Probabilistic Safety Assessment and Management Conference*, pp. 1401–1408, Research Publishing Services, 2020.

[9] A. Blanco-M., P. Marti-Puig, K. Gibert, J. Cusidó, and J. Solé-Casals, "A Text-Mining Approach to Assess the Failure Condition of Wind Turbines Using Maintenance Service History," *Energies*, vol. 12, no. 10, p. 1982, 2019.

[10] Q. Wang and C. Li, "Incident detection and classification in renewable energy news using pre-trained language models on deep neural networks," *Journal of Computational Methods in Sciences and Engineering*, vol. 22, no. 1, pp. 57–76, 2022.

[11] K. Katterbauer, "An integrated NLP Big Data framework for enhancing wind turbine reliability." `https://www.researchgate.net/publication/349708490_An_integrated_NLP_Big_Data_framework_for_enhancing_wind_turbine_reliability`, 2021.

[12] D. Küçük and Y. Arslan, "Semi-automatic construction of a domain ontology for wind energy using wikipedia articles," *Renewable Energy*, vol. 62, pp. 484–489, 2014.

[13] G. Ertek, X. Chi, A. N. Zhang, and S. Asian, "Text mining analysis of wind turbine accidents: An ontology-based framework," in *2017 IEEE International Conference on Big Data*, pp. 3233–3241, IEEE, 2017.

[14] G. Ertek and L. Kailas, "Analyzing a decade of wind turbine accident news with topic modeling," *Sustainability*, vol. 13, no. 22, p. 12757, 2021.

[15] G. L. Cash and M. Hatamian, "Optical character recognition by the method of moments," *Computer Vision, Graphics, and Image Processing*, vol. 39, no. 3, pp. 291–310, 1987.

[16] V. Govindan and A. Shivaprasad, "Character recognition — a review," *Pattern Recognition*, vol. 23, no. 7, pp. 671–683, 1990.

[17] S. Mori, H. Nishida, and H. Yamada, *Optical character recognition*. A Wiley-Interscience publication, New York and Weinheim: Wiley, 1999.

[18] D. Lopresti, "Optical character recognition errors and their effects on natural language processing," *International Journal on Document Analysis and Recognition (IJDAR)*, vol. 12, no. 3, pp. 141–151, 2009.

[19] ABBYY, "`https://www.abbyy.com`," 20.10.2021.

[20] Kofax, "`https://www.kofax.com/Products/omnipage`," 20.10.2021.

[21] Tesseract, "`https://github.com/tesseract-ocr/tesseract`," 20.10.2021.

[22] C. Patel, A. Patel, and D. Patel, "Optical character recognition by open source ocr tool tesseract: A case study," *International Journal of Computer Applications*, vol. 55, no. 10, pp. 50–56, 2012.

[23] R. Grishman, "Information extraction: Techniques and challenges," in *Information Extraction* (M. T. Pazienza, ed.), SpringerLink Bücher, pp. 10–27, Berlin, Heidelberg: Springer Berlin Heidelberg, 1997.

[24] D. Jurafsky and J. H. Martin, "Speech and language processing (draft)," 2020.

[25] D. Jurafsky, *Speech and language processing*. New York: Prentice Hall, 2rev. ed. ed., 2007.

[26] M. T. Pazienza, ed., *Information Extraction A Multidisciplinary Approach to an Emerging Information Technology*. Lecture Notes in Computer Science, Berlin, Heidelberg: Springer Berlin Heidelberg, 1997.

[27] S. G. Small and L. Medsker, "Review of information extraction technologies and applications," *Neural Computing and Applications*, vol. 25, no. 3-4, pp. 533–548, 2014.

[28] J. Oevermann, "Semantic pdf segmentation for legacy documents in technical documentation," *Procedia Computer Science*, vol. 137, pp. 55–65, 2018.

[29] H. Chao and J. Fan, "Layout and content extraction for pdf documents," in *Document Analysis Systems VI* (D. Hutchison, T. Kanade, J. Kittler, J. M. Kleinberg, F. Mattern, J. C. Mitchell, M. Naor, O. Nierstrasz, C. Pandu Rangan, B. Steffen, M. Sudan, D. Terzopoulos, D. Tygar, M. Y. Vardi, G. Weikum, S. Marinai, and A. R. Dengel, eds.), vol. 3163 of *Lecture Notes in Computer Science*, pp. 213–224, Berlin, Heidelberg:

Springer Berlin Heidelberg, 2004.

[30] T. Joachims, "Text categorization with support vector machines: Learning with many relevant features," in *European conference on machine learning*, pp. 137–142, Springer, Berlin, Heidelberg, 1998.

[31] E. Leopold and J. Kindermann, "Text categorization with support vector machines. how to represent texts in input space?," *Machine Learning*, vol. 46, no. 1/3, pp. 423–444, 2002.

[32] T. Mikolov, I. Sutskever, K. Chen, G. S. Corrado, and J. Dean, "Distributed representations of words and phrases and their compositionality," in *Advances in Neural Information Processing Systems* (C. J. C. Burges, L. Bottou, M. Welling, Z. Ghahramani, and K. Q. Weinberger, eds.), vol. 26, Curran Associates, Inc., 2013.

[33] J. Devlin, M.-W. Chang, K. Lee, and K. Toutanova, "BERT: Pre-training of Deep Bidirectional Transformers for Language Understanding," in *Proceedings of the 2019 Conference of the North American Chapter of the Association for Computational Linguistics*, pp. 4171–4186, 2019.

[34] W.-C. Chang, H.-F. Yu, K. Zhong, Y. Yang, and I. Dhillon, "X-bert: extreme multi-label text classification with using bidirectional encoder representations from transformers," *arXiv preprint arXiv:1905.02331*, 2019.

[35] Z. Zhang, J. Liu, and N. Razavian, "Bert-xml: Large scale automated icd coding using bert pretraining," *arXiv preprint arXiv:2006.03685*, 2020.

[36] S. Pfaffel, S. Faulstich, and K. Rohrig, "Performance and reliability of wind turbines: A review," *Energies*, vol. 10, no. 11, p. 1904, 2017.

[37] K. Leahy, C. Gallagher, P. O'Donovan, and D. T. J. O'Sullivan, "Issues with data quality for wind turbine condition monitoring and reliability analyses," *Energies*, vol. 12, no. 2, p. 201, 2019.

[38] Wilkinson, M. and Hendriks, B. and Spinato, F. and Gomez, E. and Bulacio, H. and Roca, J. and Tavner, P. and Feng, Y. and Long, H., "Methodology and results of the reliawind reliability field study," in *European Wind Energy Conference*, vol. 3, pp. 1984–2005, 2010.

[39] Fördergesellschaft Windenergie und andere Erneuerbare Energien, "Technical Guidelines for Power Generating Units – State-Event-Cause code for power generating units (ZEUS)," 2013.

[40] K. Kowsari, K. J. Meimandi, M. Heidarysafa, S. Mendu, L. Barnes, and D. Brown, "Text Classification Algorithms: A Survey," *Information*, vol. 10, no. 4, p. 150, 2019.

Investigation and standardisation of electrolysers for green hydrogen production using the test infrastructure in Bremerhaven, Germany

Lorenz Beck, Jannik Barthel, Gesa Quistorf, Kevin Schalk, Nora Denecke

Fraunhofer-Institute for Wind Energy Systems, 27572 Bremerhaven, Germany

Abstract. To realise the installation targets of 10 GW electrolysis capacity in Germany and 40 GW in the EU until 2030, extensive standardisation and testing is required. The electrical characteristics of electrolysers need to be fully understood and mastered for a reliable operation within distribution grids and decoupled grids of systems to produce green hydrogen, e.g. offshore. Using a modular test infrastructure and multi-functional test pads, electrolysers, but also fuel cells and CHP units, can be tested in emulated grid conditions. To emphasize the range of feasibility regarding testing, three different scenarios were developed, covering the challenges between single unit grid connection and offshore hydrogen production. The test results serve as input for standardisation committees to expand existing and develop new guidelines.

1. Introduction

Green hydrogen will play a major and decisive role in the future as it can be used for a wide range of applications in the industry, mobility, and energy sector. An amount of up to 10 million tons will be needed by 2030 in the European Union [1]. Hydrogen industry has developed an ambitious plan to reach two times 40 GW of electrolysis capacity by 2030, installed in the EU and EU neighbour countries [1]. Moreover, almost all EU member states have included plans for clean hydrogen in their National Energy and Climate Plans. Thus, the government of Germany published the coalition agreement of 2021 with the ambition to establish 10 GW of electrolysis capacity by 2030 [2].

However, the fulfilment of these ambitions is threatened if green hydrogen production technology does not enhance regarding fundamental technical challenges. Electrolysers are currently to be installed in modules in the low MW range up to 10 MW per plant [3]. Considering the current plant size, but also taking future plant capacities of 100 MW into account the electrical characteristics of such systems in terms of grid compatibility need to be fully understood [4, 5]. Since electrolysers are nonlinear, dynamic loads connected to the grid via power electronics it can be assumed that grid perturbations such as harmonics will be caused and thus, an in-depth analysis and controller design is required [6–8]. An increasing electrolysis capacity connected to the European grid will exacerbate the effect [7, 9].

The direct coupling of wind energy and electrolysis promises vast potential in production capacity and reduction of the levelised costs of hydrogen [10]. The electrical grid of such plants would be completely converter-based, without rotating generators, presenting a novel challenge in energy systems research [11]. Grid stability and the capability to black start are only a few aspects relevant for the reliable operation of decoupled energy systems for hydrogen production [12].

This paper describes the challenges of hydrogen production regarding the electrical requirements and presents a research infrastructure capable to find corresponding solutions. The first section gives an insight into state-of-the-art of standardisation and the status of hydrogen production energy systems with regard to grid requirements. The research infrastructure of the Fraunhofer-Institute for Wind Energy Systems (IWES) is presented in this context with its capability to test and validate complex grid scenarios.

2. State of standardisation and approaches to further development

The hydrogen industry currently experiences a boost equally in market demand and in the development of large-scale systems [4]. Consequently, the industry itself but also regulation authorities and technical committees recognise the need to develop the applying standards according to new technical challenges [13, 14]. The status in connection and grid-relevant standardisation shall be briefly described in this section to emphasise the demand for new test and validation methods for electrolysers.

The Transmission System Operators (TSOs) and Distribution System Operators (DSOs) are obliged to ensure a reliable and safe grid operation [15]. For this reason, the European Network of Transmission System Operators for Electricity (ENTSOE) issues grid codes to regulate and harmonise standardisation in Europe. The EU 2016/1388 Demand Connection Code (DCC) [16] is a framework of requirements for demand facilities to connect to the grid. Technical committees on international, European (e.g. ISO / IEC, CENELEC) and national level (e.g. Verband der Elektrotechnik VDE) thereupon publish standards and test specifications applicable and valid for the industry. The following figure 1 depicts the grid-relevant regulations and guidelines for systems of renewable production and hydrogen production.

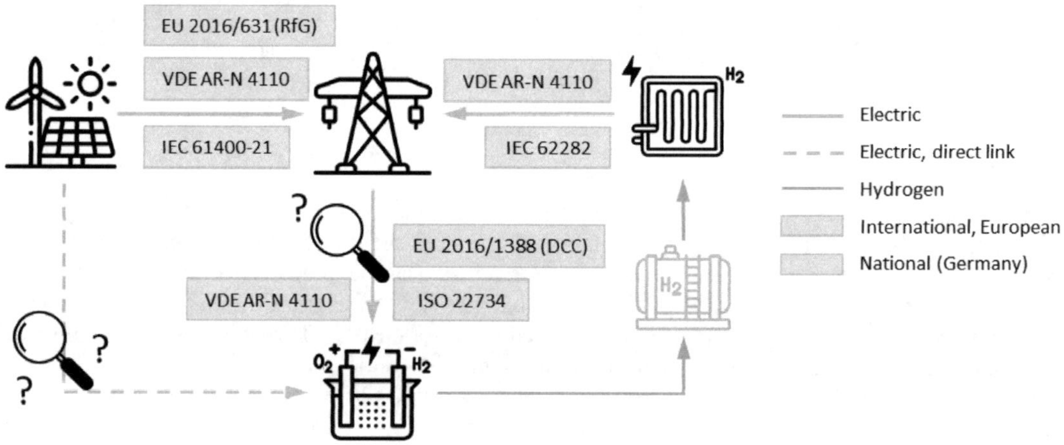

Figure 1. Relevant standards and guidelines for systems of renewable energy and hydrogen production.

Electrolyser manufacturers aim to design their products in accordance with applicable standards, e.g. the ISO 22734 (Hydrogen generators using water electrolysis - Industrial, commercial, and residential applications) [17]. Regarding grid connection, however, the EU regulation (DCC) and also the standardisation by technical committees, such as the VDE AR – N 4110 (Technical Connection Rules for Medium-Voltage) do not cover the arising issues caused by increasing power electronics penetration [18, 19], yet. The grid balancing inertia provided by synchronous machines will decrease and can partially be compensated by electrolysers. However, the electrical capabilities to provide for example Frequency Containment Reserve need to be tested and proven. Currently, working groups and round tables, to mention the European Clean Hydrogen Alliance [13] and the H2Giga [14] working group on standardisation, elaborate the new requirements with the objective to push the guidelines forward.

The ongoing demonstrator projects ERM Dolphyn [20] in the United Kingdom and H2Mare [21] in Germany explore the technical feasibility of offshore hydrogen production, coupling wind energy with electrolysis. Methods will be developed to test and validate the technical requirements needed for operation. Fraunhofer IWES contributes to the development in standardisation and to the realisation of the H2Mare demonstrator project with the unique research infrastructure, which will be presented in the following section.

3. Test benches at Fraunhofer IWES in Bremerhaven

The Fraunhofer IWES in total offers three hydrogen labs in Germany. One each in Bremerhaven, Leuna and Görlitz. This paper focuses on the test facilities in Bremerhaven offering different test benches designed for wind turbines and hydrogen systems. The existing test benches Dynamic Nacelle Testing Laboratory (DyNaLab) and Hardware-in-the-Loop Grid Compliance Test Bench (HiL-GridCoP) are designed to test the mechanical and electrical systems of wind turbines. The Hydrogen Lab Bremerhaven (HLB) consists of different hydrogen-based power generating units (PGU) and power consumption units (PCU). The special benefit of the test benches is that they all can be connected to the Power Electronics Grid Simulator (PEGS, located at the DyNaLab). The mentioned systems are described in more detail in the following chapters. Figure 2 shows an overview of Bremerhaven with focus on the test benches.

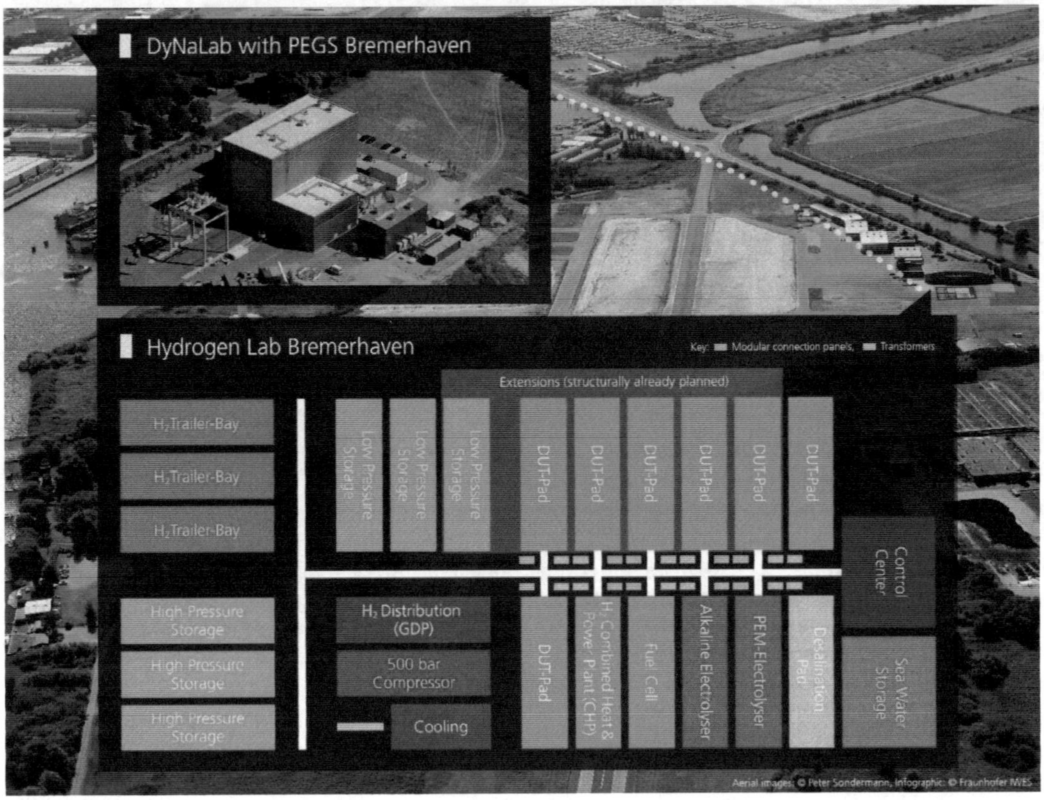

Figure 2. Overview of Bremerhaven with the DyNaLab and Hil-GridCoP test benches in the upper left and the HLB in the middle right, being connected via a medium voltage and communication cable.

Figure 2 shows the electrical test facilities DyNaLab, Hil-GridCoP and HLB in Bremerhaven. They are all in direct geographical proximity to each other and connected via a medium voltage and a communication cable. The connection of the test benches offers the opportunity to connect different units with each other and to perform specific tests.

3.1. Hydrogen Lab Bremerhaven (HLB)

The HLB combines hydrogen generating and consuming units on a test field with focus on the test, validation, and standardisation of the electrical aspects for both the overall system and the single components. The lab is designed to investigate in particular the systemic characteristics of electrolysers and its subsystems powered by renewable energies to produce green hydrogen. As shown in figure 2 the HLB consists of various PGUs and PCUs. A list of the main components is shown in table 1.

Table 1. List of units at the HLB.

Unit	Nominal power	Block colour in **Figure *2***	Description
PEM Electrolyser	1.4 MW	Blue	PEM electrolyser with nine stacks and nominal H_2 production rate of 210 m_N^3/h. Heat extraction is provided for seawater desalination.
Alkaline Electrolyser	1.3 MW	Blue	Alkaline electrolyser with three stacks and nominal H_2 production rate of 270 m_N^3/h.
Fuel Cell	270 kW	Orange	The fuel cell converts hydrogen back into electrical energy, which can be fed into the grid.
CHP	170 kW_{el} 183 kW_{th}	Orange	The combined heat and power unit (CHP) contains a hydrogen powered generator to supply electrical energy and heat.
DUT-Pads	5.0 MW	Green (bright)	The device under test (DUT) pads can flexibly be used to connect all kinds of PGUs and PCUs.

Table 1 shows the nominal power of the units, the block colour with respect to figure 2 and further a short explanation. The proton-exchange membrane (PEM) and alkaline electrolyser are power consuming, the fuel cell and combined heat and power unit (CHP) are power generating units. The backbone of the HLB consists of the downstream hydrogen infrastructure including the low- and high-pressure gas storage, gas compression, gas analytics and a medium voltage power grid.
The device under test (DUT) pad is designed as open platform for electrolysers, fuel cells, CHP units and electrical components. The nominal power of a single DUT pad is 5 MW. A DUT can easily be connected to the further infrastructure of Fraunhofer IWES. A connection panel offers a n interface for all required media, electric and data connections. Peripheral facilities for gas compression and storage are part of the Hydrogen Lab enabling a simple plug-and-play test procedure. The following figure 3 shows the concept.

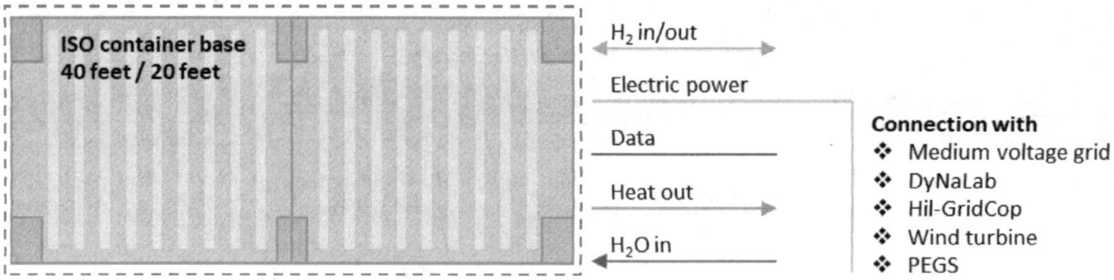

Figure 3. Design and concept of the DUT pad

3.2. Dynamic Nacelle Testing Laboratory (DyNaLab)
The DyNaLab is a nacelle test bench with a realistic testing environment for wind turbines. The DyNaLab consists of two converter controlled synchronous motors to drive the shaft of a wind turbine installed on the test bench. The motors are arranged in tandem and have a drive power of 5 MW each. This means that 10 MW (up to 15 MW in overload condition) are provided for the test operations. The engines are directly coupled to the high-speed shaft of the nacelle and are used to reproduce realistic generator torques using a Hardware-in-the-Loop (HiL) technique. The electrical side of the nacelle can be connected to the public grid or to the PEGS, which emulates an electrical grid and is described in more detail in Section 3.5. At the DyNaLab both electrical and mechanical tests can be performed [22].

3.3. Hardware-in-the-Loop Grid Compliance Test Bench (Hil-GridCoP)
Hil-GridCoP is the second wind turbine test bench in Bremerhaven. In contrast to the DyNaLab, the Hil-GridCoP test bench will only be used to test the minimal system comprising of a high-speed generator, a converter, and a transformer. The converter controlled synchronous motor has a nominal power of 9 MW (up to 13 MW in overload condition) and uses HiL technique for reproducing realistic generator torques. The electrical side of the DUT can be connected to the public grid or to the PEGS analogously to the DyNaLab. At Hil-GridCoP, tests for electrical properties are performed [23].

3.4. Local research wind turbine
Since 2017, the Fraunhofer IWES has been using a wind turbine installed in immediate vicinity of the HLB for research purposes. This wind turbine will provide the renewable energy to perform electrolysis with nearly zero emissions. Furthermore, the wind turbine can be connected to the same grid as the test benches. This allows further extensive test scenarios to be implemented.

3.5. Power Electronics Grid Simulator (PEGS)
The PEGS is a voltage controlled high performance converter system. PEGS is used to emulate an artificial medium voltage grid which enables the possibility to perform static and dynamic tests. Within the static tests e.g., the determination of the effective and reactive power outputs for various grid conditions, can be examined. Furthermore, within the dynamic tests transient grid events like voltage dips, frequency changes, and phase angle jumps can be emulated. The functionality of PEGS can be used for validation and research purposes for the DUTs on the test benches [24].

Summary of characteristic values of the PEGS:

- 10/20/36 kV nominal voltage levels
- Frequency range from 45 to 65 Hz
- 44 MVA installed converter power
- < 3 % Total Harmonic Distortion (THD) at 50 Hz

Possible grid event simulations by PEGS:

- Frequency changes from 45 to 65 Hz with a ramp up to 20 Hz/s
- Phase angle changes from 0 to 180 ° with a ramp up to 90 °/ms
- Symmetrical and unsymmetrical faults
- Under voltage and over voltage ride through emulation
- Emulation of virtual impedances

4. The combined infrastructure and feasible test scenarios
The particular benefit of the test benches described above is that they can all be connected to the same medium voltage system. This means each unit individually or collectively can be connected to the PEGS and thus all electrical tests can be performed with each unit. To introduce the possibilities offered by this test infrastructure, three different test scenarios are described below. Figure 4 shows a comprehensive illustration of the overall system.

Journal of Physics: Conference Series 2257 (2022) 012005 doi:10.1088/1742-6596/2257/1/012005

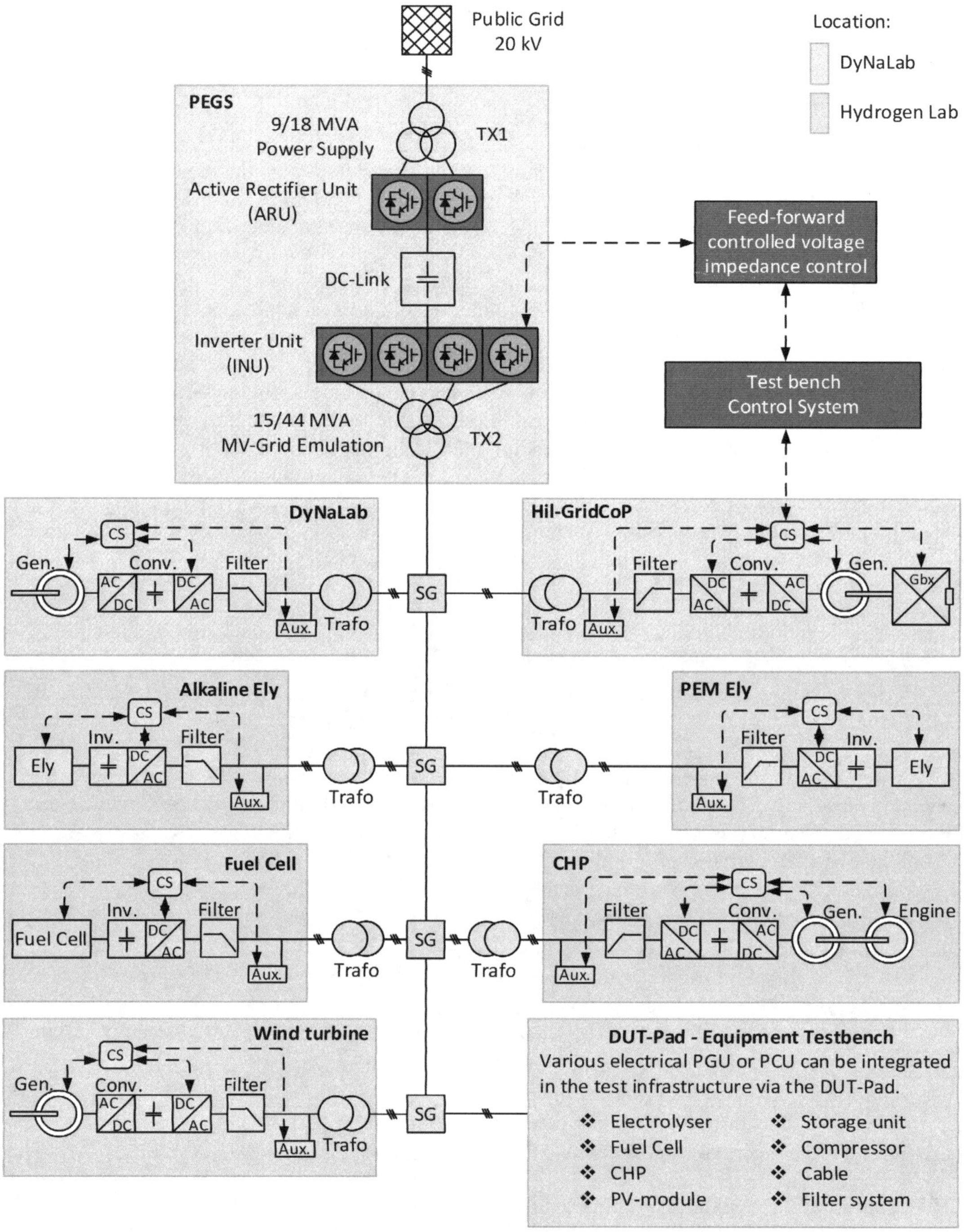

Aux. = Auxialliary
CS = Control System
Gbx. = Gearbox
PCU = Power consumption unit
PV = Photovoltaik
CHP = Combined Heat and Power unit
DUT = Device under Testing
Gen. = Generator
PEGS = Power Electronics Grid Simulator
SG = Switchgear
Conv. = Converter
Ely = Electrolyser
Inv. = Inverter
PGU = Power generating unit

Figure 4. Overall view of the electrical test infrastructure from Fraunhofer IWES in Bremerhaven

Figure 4 shows the overall test infrastructure from Fraunhofer IWES in Bremerhaven, including the test benches described in chapter 3. A superordinated control system allows to orchestrate the system. The system's grid can be decoupled from the public grid enabling a safe test environment. The connection between the test benches is realised via a medium voltage system consisting of cables and switchgears. The whole system can be either connected to the public grid of the local DSO or to the emulated voltage of the PEGS. The possibility of unrestricted connecting and disconnecting of the test benches offers various configurations. On the one hand it is possible to test only one PGU or PCU at a time and on the other hand multiple unit tests with up to seven or more devices under testing at the same time are possible. As an overview three different test scenarios are described below to emphasise the range of feasibility.

4.1. Test scenario one – Single unit validation

The test constellation is a single DUT connected to the PEGS. Hereby, it is irrelevant whether the DUT is a PGU or a PCU. The objective is to test the DUT in accordance with the requirements of technical guidelines or valid grid codes. Therefore, the determination of the electrical properties is necessary to verify the grid compliance. The results will also be used for the further development of standards and guidelines. Beyond that, the DUT-models can be validated by comparing the test results with the simulation.

4.2. Test scenario two – Offshore hydrogen production

The test constellation emulates an island grid with a single wind turbine and a single electrolyser. This configuration can be found in a greater extent in offshore wind power plants with a direct electrolyser coupling. Different research projects focus on the investigation of this scenarios within the next years [20, 21]. Up to 5 MW of electrolysis capacity emulated together with an offshore wind turbine generation profile can be investigated. Objective of this scenario could be the evaluation of the technical availability and efficiency of an offshore hydrogen production, as the interactions of the different systems are almost unexplored.

In addition to the electrical aspects, the supply of feed water is a crucial technical challenge regarding system efficiency and environmental issues. To research and optimise the process, a seawater reservoir and a test pad for desalination units are integrated in the HLB. The thermal energy of the PEM electrolysis is extracted and used for desalination processes.

4.3. Test scenario three – Microgrid of wind and hydrogen units

The third test scenario describes a microgrid consisting of different types of PGUs and PCUs. This microgrid forms a comprehensive network with seven or more DUTs. The advantage of this configuration is the possibility to study the control technology and the interaction between the individual units. By connecting and disconnecting DUTs, a conclusion can be drawn about the effect of a single device on the overall system. In this scenario, the volatile behaviour of generators and consumers can be extensively tested, real-world grid conditions can be emulated and the interacting behaviour between the various components can be evaluated

5. Conclusions

To realise the electrolyser installation targets of 10 GW in Germany and 40 GW in the EU until 2030, extensive standardisation and testing is required. There is a lack of standards and technical guidelines for the process to connect hydrogen production facilities to power grids, being a hurdle for large-scale implementation. The electrical characteristics of electrolysers need to be fully understood and mastered for reliable operation of distribution grids and the decoupled grids of systems to produce green hydrogen, e.g. offshore.

With the Hydrogen Lab Bremerhaven, the research infrastructure of the Fraunhofer IWES offers the up to this point worldwide unique possibility to test and validate hydrogen equipment in a realistic and reliable environment before deploying them into the field. On the basis of a modular test

infrastructure and multi-functional test pads, electrolysers, but also fuel cells and CHP units, can be tested in emulated grid conditions.
To emphasise the range of feasibility regarding testing, three different scenarios were developed. The first scenario *Single unit validation* describes one single DUT to be connected to the electrical grid. This is the current standard in validation and certification processes. The second scenario *Offshore hydrogen production* describes an island grid with a single wind turbine and a single electrolyser. This configuration represents future offshore wind power plants with direct electrolyser coupling. The third scenario *Microgrid of wind and hydrogen units* describes a network with seven or more DUTs consisting of different types of PGUs and PCUs.
The results serve as input standardisation committees to develop new guidelines. With more than five years of experience with wind energy testing methods, Fraunhofer IWES is looking forward transferring this knowledge to hydrogen units.

Acknowledgments
This work is supported by the German Federal Ministry for Economic Affairs and Climate Action, the European Regional Development Fund, and the Federal State of Bremen.

References

[1] EUROPEAN COMMISSION 2020 *A hydrogen strategy for a climate-neutral Europe* (Brussels)
[2] SPD, BÜNDNIS 90/DIE GRÜNEN UND FDP 2021 *Mehr Fortschritt wagen: Bündnis für Freiheit, Gerechtigkeit und Nachhaltigkeit* (Berlin)
[3] sunfire GmbH *Sunfire-HyLink Alkaline: Renewable hydrogen for all applications*
[4] sunfire GmbH 2021 *Bereit für Wasserstoff-Großprojekte: Sunfire erhöht die Fertigungskapazität für Alkali-Elektrolysure bis 2023 auf 500 MW/Jahr: Pressemitteilung* (Dresden)
[5] IEA 2019 *The Future of Hydrogen* (Paris)
[6] Khan A, Hosseinzadehtaher M, Shadmand M B, Bayhan S and Abu-Rub H 2020 *EEE Ind. Electron. Mag.* **14** 65–78
[7] MIGRATE – Massive InteGRATion of power Electronic devices 2016 *Deliverable D1.1 Report on systemic issues*
[8] Hans F 2021 Input Admittance Modeling and Passivity-Based Stabilization of Digitally Current-Controlled Grid-Connected Converters Universitätsbibliothek Braunschweig
[9] Divan D and Kandula P 2016 *CPSS TPEA* **1** 57–65
[10] KPMG International 2020 *The hydrogen trajectory*
[11] MIGRATE – Massive InteGRATion of power Electronic devices 2016 *Deliverable 3.1 Description of system needs and test cases*
[12] Castillo A 2013 Microgrid provision of blackstart in disaster recovery for power system restoration *2013 IEEE International Conference on Smart Grid Communications (SmartGridComm) 2013 IEEE International Conference on Smart Grid Communications (SmartGridComm) (Vancouver, BC, Canada, 21.10.2013 - 24.10.2013)* (IEEE) pp 534–9
[13] THE EUROPEAN CLEAN HYDROGEN ALLIANCE 2021 *REPORTS OF THE ALLIANCE ROUNDTABLES ON BARRIERS AND MITIGATION MEASURES*
[14] DBI Gruppe 23.12.21 *Verbundvorhaben H2Giga - Technologieplattform Elektrolyse: Teilvorhaben DBI: Abbau von Innovationshürden*
[15] THE EUROPEAN COMMISSION 2017 *Commission Regulation (EU) 2017/1485 of 2 August 2017 establishing a guideline on electricity transmission system operation*
[16] THE EUROPEAN COMMISSION 2016 *Commission Regulation (EU) 2016/1388 of 17 August 2016 establishing a Network Code on Demand Connection*
[17] International Organization for Standardization 2019 Hydrogen generators using water electrolysis — Industrial, commercial, and residential applications *(*ISO 22734:2019*)*

[18] Marlon Koralewicz, Josephine Glandien, Jens Hüttenrauch, Anja Wehling, Barbara Fricke, Felix Künkel, Clemens Heitsch, Dr. Manuela Jopen, Myriam Panofen and Chi Yan Tang 2020 *PORTAL GREEN: Power-to-Gas-Leitfaden zur Integration Erneuerbarer Energien*

[19] VDE Verband der Elektrotechnik Elektronik Informationstechnik e. V. 2018 Technical Connection Rules for Medium-Voltage *(*VDE-AR-N 4110*)*

[20] David Caine, Molly Iliffe, Kevin Kinsella, Widya Wahyuni, Laura Bond 2019 *Dolphyn Hydrogen - Large Offshore Hydrogen Production: Phase 1 - Final Report* (ERM Worldwide Group Ltd Department for Business, Energy and Industrial Strategy)

[21] Bundesministerium für Bildung und Forschung *H2MARE: Wie Partner im Leitprojekt H2Mare Wasserstoff direkt auf hoher See produzieren wollen*

[22] Neshati M, Zuga A, Jersch T and Wenske J 2016 Hardware-in-the-loop drive train control for realistic emulation of rotor torque in a full-scale wind turbine nacelle test rig *2016 European Control Conference (ECC) 2016 European Control Conference (ECC) (Aalborg, 29.06.2016 - 01.07.2016)* (IEEE) pp 1481–6

[23] Christian Mehler, Torben Jersch, Jan Wenske and Britta Rollert 2018 *CERTIFICATION OF ELECTRICAL CHARACTERISTICS ON SYSTEM TEST BENCHES: Speeding up Time to Market with Testing for Grid Integration* (Fraunhofer IWES)

[24] Azarian S, Jersch T and Khan S 2019 Comparison of impedance behavior of UVRT-Container with medium voltage grid simulator in case of unsymmetrical voltage dip *2019 Conference for Wind Power Drives (CWD) 2019 Conference for Wind Power Drives (CWD) (Aachen, Germany, 12.03.2019 - 13.03.2019)* (IEEE) pp 1–12

Investigation of manufacturing-related deviations of the bearing clearance on the performance of a conical plain bearing for the application as main bearing in a wind turbine

Amadeus Rolink(*), Georg Jacobs, Matthias Müller, Timm Jakobs, Dennis Bosse

Chair for Wind Power Drives (CWD)
Campus-Boulevard 61
52074 Aachen, Germany

amadeus.rolink@cwd.rwth-aachen.de

Abstract. The use of a segmented conical plain bearing for the application as main bearing in a wind turbine offers multiple advantages. The main bearing can be designed very compact. In addition, in case of a main bearing damage, individual bearing segments can be exchanged on the tower without the use of an external crane, in contrast to the repair procedure of rolling bearings. It is well known from conventional applications of plain bearings that the bearing clearance has a major influence on the load carrying capacity. Therefor the realization of narrow tolerances in the manufacturing of the bearing components is necessary. In this paper, the influence of the manufacturing tolerances on the performance of conical plain bearings is investigated. The influence of the tolerances on the load carrying capacity can be determined by conducting elastohydrodynamic (EHD) simulations. Finally, an evaluation of the robustness of the conical bearing concept against manufacturing-related deviations is carried out.

1. Introduction and objective

In recent years, plain bearings have got in the focus of research for the application as main bearing in a wind turbine (WT). These type of bearings can be used as an alternative to the currently used rolling bearings that often fail before their designed life time [1,2]. In comparison to roller bearings, these bearings offer the great advantage that they can be designed in segments, thus enabling segment exchange on the tower in the case of a bearing damage. This results in significantly lower repair costs and ultimately in lower levelized costs of energy (LCOE).

In this context, during the research project WEA-GLiTS, a plain bearing prototype as moment bearing with conical, tapered sliding surfaces (see Figure 1, left) and an elastic segment support structure was developed and successfully tested [3,4]. The performed tests on a full-size WT test bench exceeded expectations regarding the performance at high dynamic loads and low rotational speeds.

A requirement for the functionality of a plain bearing is the existence of a convergent lubrication gap between the bearing surface and the rotating shaft [5]. For the bearing prototype this convergent lubrication gap (~5/10 mm) is several orders of magnitude smaller than the bearing diameter (~1 m). In the context of a continuing development of this concept with the aim of an industrial application, the influence of the manufacturing tolerances on the lubrication gap and finally on the load carrying capacity of the bearing needs to be investigated. Secondly, the objective of this paper is the identification of the necessary manufacturing tolerances to ensure a proper pressure distribution, since the manufacturing precision directly affects the manufacturing costs and finally the LCOE.

2. Approach

When manufacturing components, deviations from the nominal geometry can never be avoided. Therefore, in the design process tolerances are specified. Within these tolerances, the dimensions of the component are allowed to vary without affecting the functionality of the component. For this purpose, ISO 2768 specifies general tolerances of the classes very coarse, coarse, medium and fine [6]. However, individual tolerances may have to be assigned to individual dimensions of the tolerance chain in order to ensure the functionality of the assembly.

Regarding plain bearings, the height of the lubrication gap is crucial for the functionality of the bearing. For the characterization of a conventional radial plain bearing, the relative bearing clearance ψ is used as the ratio of the nominal lubrication gap height S to the bearing radius:

$$\psi = \frac{R_{Bearing} - R_{shaft}}{R_{Bearing}} = \frac{S}{R_{Bearing}} \quad (1)$$

In the case of a plain bearing with conical sliding surfaces that is considered in this paper, the relative bearing clearance is defined at the mean radius R_m of the cone (see Figure 1).

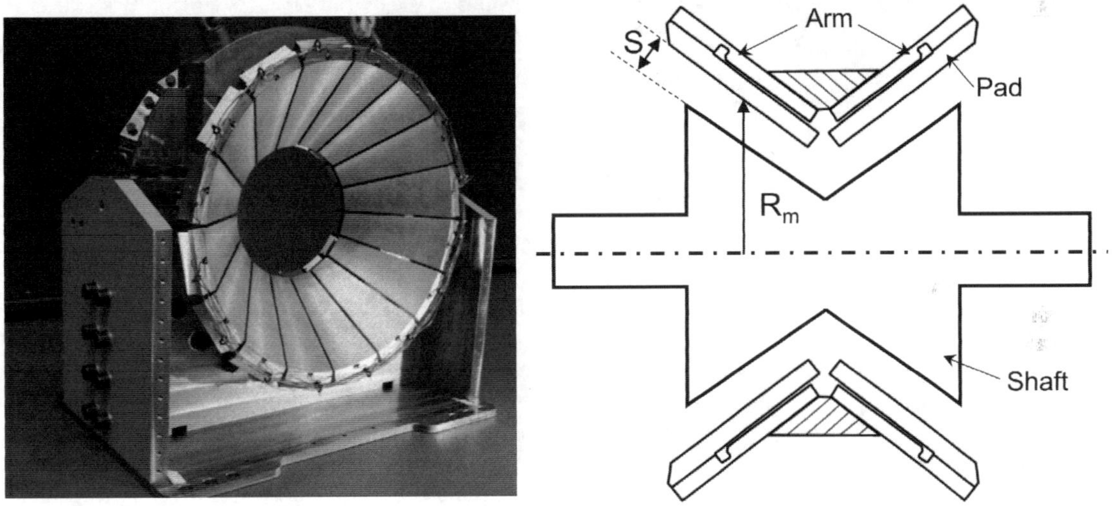

Figure 1. Conical plain bearing demonstrator (left) and definition of relative bearing clearance (right, increased representation)

$$\psi_{conical} = \frac{S}{R_m} \quad (2)$$

In addition, Figure 1 (right) shows the flexible support structure of the bearing pads. The pads are connected to the housing by flexible arms that allow the pads to align to the shaft under load. This enables a uniform pressure distribution and prevents edge loading.

In the first step of the investigation, the influence of the tolerance classes (ISO 2768 fine and individual tolerances) on the bearing clearance is investigated by using an approach according to Fischer [7]. This approach can be used for a two-dimensional tolerance stackup analysis. For a three-dimensional tolerance stackup analysis, an additional computer software would be required.

The analysis will be done for the design of a scaled test bearing (outer diameter ~500 mm) that will be tested on a test bench at the Chair for Wind Power Drives (CWD). The scaling, which is assuring a constant Sommerfeld number, is necessary for cost reasons, since different bearing geometries are tested within the project. Subsequently, elastohydrodynamic (EHD) simulations are performed to evaluate the effect of the bearing clearance variations on the hydrodynamic pressure distribution.

3. Tolerance stackup analysis

As part of the research project FlexPad, funded by the German Federal Ministry for Economic Affairs and Climate Actions, experimental investigations for this conical bearing type are planned on a test bench at the CWD. To be able to test different bearing variants in a time- and cost-efficient manner, the individual bearing variants are assembled using a modular system (see Figure 2).

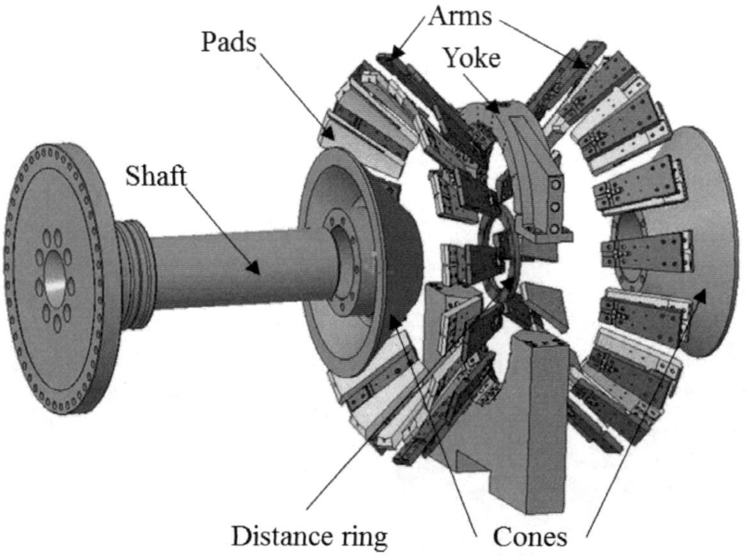

Figure 2. Exploded view of conical bearing variant

The rotating part of the assembly consists of a shaft onto which two cones are mounted. These two cones are axially positioned by a distance ring. The pads of the test bearing are bolted to the flexible arms, which are mounted to the upper and lower yoke. The question that must be answered before carrying out the tests, and which is also important for the industrialization of such a concept, is what influence the manufacturing tolerances have on the bearing clearance. For this purpose, a tolerance stackup analysis according to Fischer is carried out [7].

The first step of such an analysis is to define for which distance the tolerance stackup analysis is performed. In this case, this is the bearing clearance as the gap between point I on the shaft and point II on the pad (see Figure 3).

Figure 3 reveals that the nominal value S_0 of the bearing clearance can vary between S_{min} and S_{max}, since the final position I and II can differ due to the tolerance deviations. To determine the values of S_{min} and S_{max}, the tolerance analysis is performed for both the cone and the pad assemblies. Therefore, the tolerance chain is set up in the horizontal (x-direction) and vertical direction (y-direction) for each assembly and the maximum tolerances are added up. Figure 4 shows this procedure as an example for the shaft assembly in vertical direction.

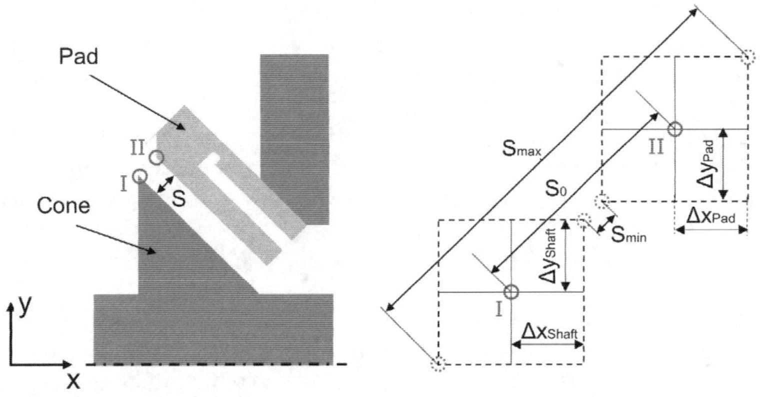

Figure 3. Illustration of the bearing clearance and possible deviation in vertical and horizontal direction

Figure 4 shows the tolerance chain of the shaft assembly in vertical direction from the reference (center line) to the target I. The individual dimensions of the tolerance chain that have an influence on the position of the point I are numbered and shown with their directions in the figure on the left and are listed in the table on the right. In addition to the nominal dimension, the corresponding maximum tolerances are listed in the table. Summed up, this results in the total tolerance deviation in vertical direction ΔY_{Shaft} (marked in yellow). One advantage of this table is that, in addition to the absolute tolerances, their share of the total tolerance is also shown. This makes it immediately clear which tolerances must be adjusted if the total tolerance is too large.

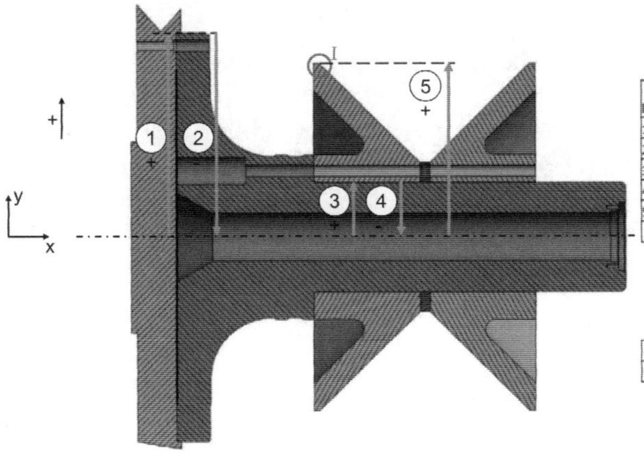

	Nominal length [mm]			
Description	[+]	[-]	Tolerance [mm]	Share [%]
1. Centering rim	298.500		0.050	30.5%
2. Centering pin		298.500	0.044	26.8%
3. Radius shaft	160.000		0.025	15.2%
4. Inner radius cone		160.000	0.025	15.2%
5. Outer radius cone	253.460		0.020	12.2%
Total	711.960	458.500	0.164	

Nominal length: 253.460 mm

	Nominal length [mm]	Tolerance [mm]	Minimum [mm]	Maximum [mm]
Arithmetical analysis (worst case)	253.460	0.164	253.296	253.624
Statistical analysis (RSS)	253.460	0.078	253.382	253.538

Figure 4. Tolerance chain for shaft assembly in vertical direction.

If the individual tolerances are added up to a total tolerance, a worst case scenario is considered. This means that the randomly distributed tolerances add up to the maximum total tolerance. Since this scenario is very unlikely, a statistical analysis is often performed instead of the worst case scenario. By assuming that the deviations are normally distributed it is more likely that the finally manufactured dimensions lie in the middle rather than at the edge of the tolerance fields. In this context, a commonly used statistical method is the root-sum-square (RSS) method. In this method, the square root is taken from the sum of the squares of the individual tolerances T_i (see Eq. 3) [7].

$$T_{RSS} = \sqrt{T_1^2 + T_2^2 + \cdots T_n^2} \quad (3)$$

The resulting tolerance of the statistical approach is generally smaller than the tolerance of the arithmetic approach (see Figure 4, 0.078 mm compared to 0.164 mm). This makes it possible to select larger individual tolerances without exceeding the function-critical limits. However, it should be noted that not every manufactured component meets these requirements, and a certain proportion of faulty components must therefore be accepted. However, this proportion is very low with the process capabilities that are common today [7,8].

4. Results

In this section, first the results of the tolerance stackup analysis are presented. Then, results of EHD simulations for the upper and lower limit of the bearing clearance are presented to investigate the influence of the tolerance deviations on the hydrodynamic pressure distribution.

4.1. Tolerance stackup analysis

Following the approach presented in Figure 4, the tolerances in horizontal and vertical direction for the shaft assembly and the bearing assembly are determined (ΔX_{Shaft}, ΔY_{Shaft}, ΔX_{Pad}, ΔY_{Pad}). With these values the total deviations (ΔS_{Shaft} and ΔS_{Pad}) and finally the minimum and maximum bearing clearances (S_{min} and S_{max}) can be calculated (see Eq. 4 and 5).

$$S_{min} = S_0 - \Delta S_{Shaft} - \Delta S_{Pad} = S_0 - \sqrt{\Delta X_{Shaft}^2 + \Delta Y_{Shaft}^2} - \sqrt{\Delta X_{Pad}^2 + \Delta Y_{Pad}^2} \quad (4)$$

$$S_{max} = S_0 + \Delta S_{Shaft} + \Delta S_{Pad} = S_0 + \sqrt{\Delta X_{Shaft}^2 + \Delta Y_{Shaft}^2} + \sqrt{\Delta X_{Pad}^2 + \Delta Y_{Pad}^2} \quad (5)$$

The tolerance stackup analysis was carried out with the general tolerances according to ISO 2768 (fine) [6] and additionally with individual finer tolerances (see Figure 4). The individual finer tolerances were defined in cooperation with the bearing manufacturer and project partner Miba Industrial Bearings. For these two tolerance classes, the results of the statistical analysis (RSS) are shown in Table 1.

Table 1. Results of tolerance stackup analysis for tolerances according to ISO 2768 (fine) and individual finer tolerances

	ISO 2768 fine	Individual Finer Tolerances
	RSS	RSS
ΔS_{Shaft} [mm]	0.482	0.083
ΔS_{Pad} [mm]	0.464	0.070
S_{min}[mm]	-0.691	0.103
S_{max} [mm]	1.202	0.408
Ψ_{min} [‰]	-3.791	**0.563**
Ψ_{max} [‰]	6.591	**2.237**

It quickly becomes evident that the general tolerances according to ISO 2768 (fine) are unsuitable for this application, since large variation in the bearing clearance occur. Even a negative bearing clearance can occur, which means that the shaft has a larger diameter than the bearing pads. In comparison, the range of the bearing clearance for the individual finer tolerances is significantly smaller. In this case the absolute gap height varies between $S_{min} = 0.103$ mm and $S_{max} = 0.408$ mm. This corresponds to a relative bearing clearance of $\psi_{min} = 0.563$ ‰ and $\psi_{max} = 2.237$ ‰. The influence of this still quite large range is analyzed in the following section.

4.2. Effect on load carrying capacity

To analyze the influence of the variation of the bearing clearance ($\psi_{min} = 0.563$ ‰ to $\psi_{max} = 2.237$ ‰) on the load carrying capacity, EHD simulations are performed for the lower and upper limit of the bearing clearance under nominal loads during power production. For these simulations the Software First developed by the project partner IST was used.

Figure 5 shows the pressure distribution as a result of the performed EHD simulations. The load patterns result from the combination of the applied non-torque loads (thrust and radial force combined with bending moments). In general, the load patterns for the two bearing clearances look quite similar and the same pads are loaded. However, the maximum pressures on the individual pads are higher for the bearing with the larger bearing clearance. The reason for this is that in case of a larger clearance the difference in the curvature of the two sliding surfaces is larger so that the surfaces of the shaft and the pads converge at a steeper angle. This leads to the fact that the pressure built up takes place in a smaller area. Since the same bearing load is carried in a smaller area, consequently higher hydrodynamic pressures occur.

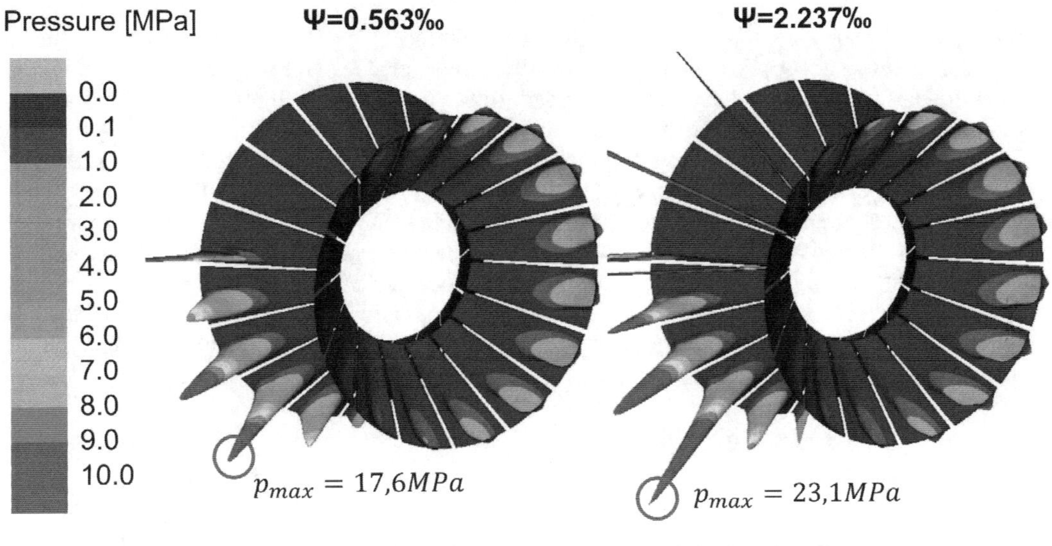

Figure 5. Pressure distributions for the lower and upper limit of the bearing clearance range.

To make this more clear, the pressure area ratio (relation of loaded area of the pad to the total area of the pad) for the front and the back cone are shown in Figure 6 and 7. It can be seen that in case of a smaller bearing clearance, the loaded zone of the front cone is larger. This becomes even more clearly in the case of the back cone. Here, in particular, the pads at the outer region of the load zone are more evenly loaded than in case of the larger bearing clearance.

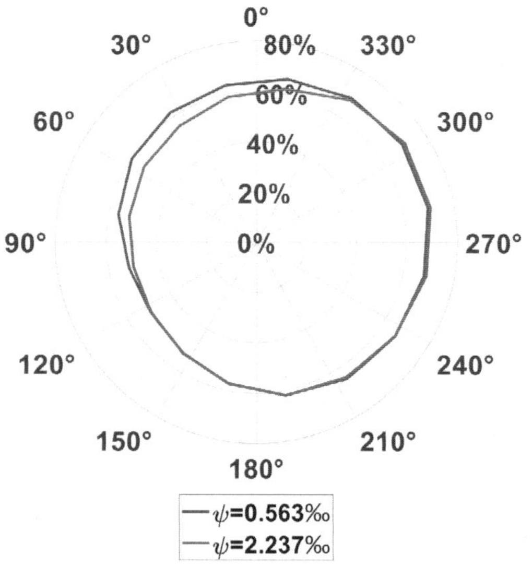

Figure 6. Loaded area of front cone for different bearing clearances

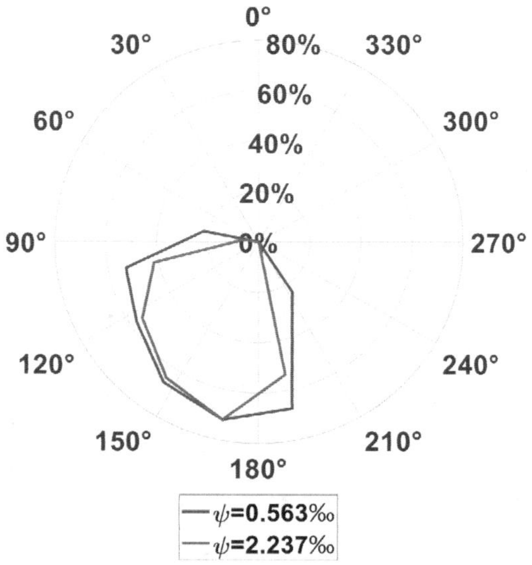

Figure 7. Loaded area of back cone for different bearing clearances

Overall, the load carrying capacity of the bearing is ensured in both cases. In the case of the larger bearing clearance, contact pressures do occur locally. However, these only exist in very small areas on the inner edge of the front cone and at the boundary of the load zone, so that after a running-in process the surfaces will adapt and the contact pressures will disappear.

The reason why this type of bearing is so robust to variations of the bearing clearances is the flexible support structure of the pads. Due to this elastic support structure (see Figure 3), the pad can align to the movement of the shaft, and thus edge loading can be avoided. In this context, Figure 8 shows the total displacement of the pads of the back cone. This plot reveals that the displacements of the pads due to the elastic deformation of the support structure (max. ~1 mm) are of the same order of magnitude as the bearing clearance (0.103 mm to 0.408 mm). Therefore, manufacturing-related deviations of the bearing clearance are not as severe for the FlexPad bearing concept as in the case of a bearing with a rigid support structure.

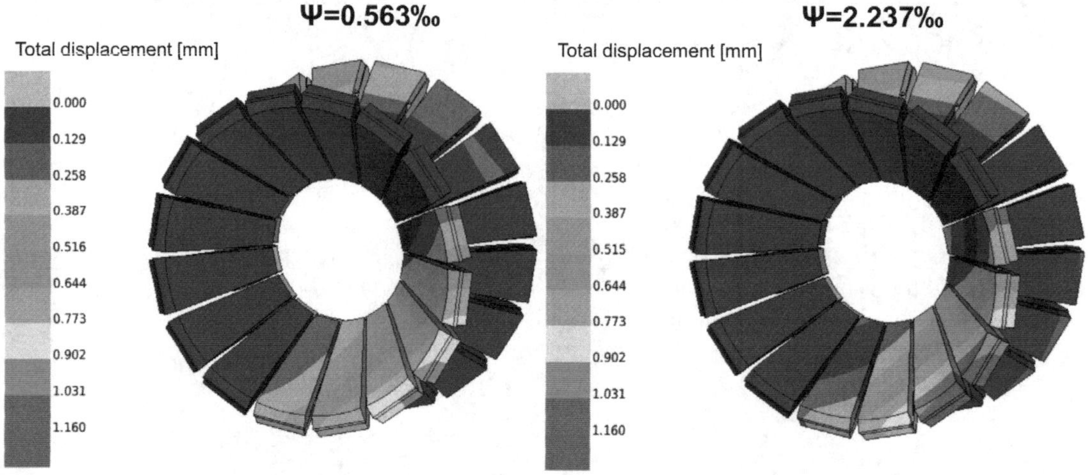

Figure 8. Total displacement of the pads of the back cone

5. Conclusion and Outlook

The objective of this investigation is to determine the influence of manufacturing-related deviations of the bearing clearance on the pressure distribution and load carrying capacity of a conical plain bearing for the application as main bearing in a wind turbine. By performing a two-dimensional tolerance stackup analysis according to Fischer [7] a range of the relative bearing clearance was determined. In this context, it became clear that the general tolerances according to ISO 2768 (fine) [6] are not applicable for the manufacturing of the bearing components and individual finer tolerances must be used to avoid direct contact.

With the help of EHD simulations, it was shown that the load carrying capacity of the bearing is ensured for the determined variation of the bearing clearance and that the FlexPad bearing concept is therefore very robust against deviations in the bearing clearance. The reason for this is the flexible support structure of the pads, which allows the pads to align to the movement of the shaft and also compensates deviations of the bearing clearance.

The results obtained in this study allow to start the manufacturing of the components for the planned experiments in the ongoing research project. The aim of these tests will be to investigate the influence of the geometrical parameters on the performance of the FlexPad bearing concept [9]. Finally, the results of this study are particularly important for a later industrial application.

6. Acknowledgment

The presented investigations are based on the results of the joint project "FlexPad – Development of a design methodology for an innovative bearing concept", promoted by the German Federal Ministry for Economic Affairs and Climate Action" (grant ID 03EE2002B). The Authors would like to thank the German Federal Ministry for Economic Affairs and Climate Action for funding the "FlexPad" project. In addition, the authors would like to thank their project partners Miba Industrial Bearings Germany GmbH (Göttingen, Germany) and IST Ingenieurgesellschaft für Strukturanalyse und Tribologie mbH (Aachen, Germany) for the excellent cooperation and support.

Supported by:
Federal Ministry for Economic Affairs and Climate Action
on the basis of a decision by the German Bundestag

References

[1] Chovan C, 2019, Seven Years of Solid Results, *Wind Systems*, March 2019, 17
[2] Fierro T, 2016, Extending Bearing Life in Wind Turbine Mainshafts, *Power Engineering*, 120
[3] Schröder T, Jacobs G, Rolink A, Bosse D, 2019, "FlexPad" – Innovative conical sliding bearing for the main shaft of wind turbines; *J. Phys.: Conf. Ser. 1222*, 12, https://doi.org/10.1088/1742-6596/1222/1/012026
[4] Schröder T, Rolink A, Jacobs G, Bosse D, 2020, Full-Size Test Bench Measurement and Validation of a Novel Conical Sliding Bearing for the Main Shaft of Wind Turbines; *Bearing World Journal* 5, (Frankfurt am Main: VDMA Verlag), pp 149-158
[5] Jacobs G, 2016, Vorlesungsumdruck Maschinengestaltung Band I, (Aachen: Druck & Verlagshaus Mainz), chapter 3 p 127
[6] ISO 2768-1: General tolerances; tolerances for linear and angular dimensions without individual tolerance indications, 1991, (Berlin: Beuth Verlag)
[7] Fischer B, 2011, Mechanical Tolerance Stackup and Analysis, (Boca Raton: CRC Press), chapter 8 pp 97-102
[8] Beucher O, 2007, Wahrscheinlichkeitsrechnung und Statistik mit MATLAB, (Berlin, Heidelberg: Springer), chapter 4 p 159
[9] Rolink A, Jacobs G, Schröder T, Keller D, Jakobs T, Bosse D, Lang J, Knoll G, 2021, Methodology for the systematic design of conical plain bearings for use as main bearings in wind turbines, *Forschung im Ingenieurswesen Band 85*, pp 629-637, https://doi.org/10.1007/s10010-021-00452-3

FLOTANT concept: floater design, integrated modelling & global performance

Jordi Serret[1], Bernardo Kahn[2], Bruce Cavanagh[2], Patricia Lorente[2], Remy Pascal[1], Clementine Girandier[2], Paul McEvoy[3], Carlos Cortés[4], Rubén Duran[5], Alejandro Romero[6]

[1]Innosea, Edinbugh, UK
[2]Innosea, Nantes, France
[3]TFI, Dublin, Ireland
[4]Esteyco, Madrid, Spain
[5]Cobra, Madrid, Spain
[6]PLOCAN, Las Palmas, Spain

jordi.serret@innosea.fr, bernardo.kahn@innosea.fr, bruce.cavanagh@abl-group.com, patricia.lorente@innosea.fr, remy.pascal@innosea.fr, clementine.girandier@innosea.fr, paul.mcevoy@technologyfromideas.com, carlos.cortes@esteyco.com, ruben.duran@grupocobra.com, alejandro.romero@plocan.eu

Abstract. This paper presents the final design of the FLOTANT concept developed within the framework of a Cooperation Research Project funded by the European Union´s Horizon 2020 research and innovation programme. The FLOTANT concept is a hybrid concrete-plastic barge-type floating offshore substructure holding a 12MW wind turbine; buoyancy is achieved using plastic material (foam) fitted within the floater substructure. The paper describes the FLOTANT project innovations and presents the principal dimensions of the floating wind turbine system. The floating system has been developed to fulfil strict requirements at two different locations: Gran Canarias (GC) and West of Barra (WoB) sites. Therefore, two different mooring solutions including innovative elastomers to reduce peak loads are presented. The floating system's intact and damaged static stability is checked along with the hydrostatic and hydrodynamic properties, showing the low-cost barge-type system's feasibility. To ensure the desired behaviour of the floating system, a relevant subset of design load cases (DLCs) based on international standard requirements have been simulated checking the global performance of floater and moorings. Fully coupled hydro-servo-aero-elastic simulations have been performed to consider the wind and wave loading, the aerodynamic response of the rotor, the hydrodynamic response and structural dynamics of the turbine-floater-mooring assembly, and the control system actions. A summary of the results of the last design loop within the FLOTANT project is presented.

1. Introduction

The FLOTANT project is developing an innovative, integrated, FOWT solution, optimised for deep waters and holding a 12MW wind turbine generator. The project is framed into a Cooperation Research Project funded by the European Union´s Horizon 2020 research and innovation programme.

The concept developed by the consortia is a hybrid concrete-plastic barge-type floating offshore substructure with a heave plate, an active ballast system, and the singularity of getting floatability by using plastic foam material fitted within the floater substructure. The concept has evolved from the beginning of the project. Initially, a hybrid spar buoy concept was considered; however, by the end of the first loop of design [1] the development was close to the actual design and the barge-type was finally developed on the basis of a cost reduction associated with the innovative construction, installation, and operations and maintenance strategies linked with such a floater type. The final design presented below includes an innovative mooring system using high-performance polymers to reduce both weight and peak loads. Another remarkable difference between the two loops of design is the inner structure of the floater. Initially, the concept received the buoyancy from plastic inflated bags fitted into the floater whereas now, the plastic bags are substituted by blocks of extruded polystyrene foam material (XPS) filling the void spaces within the floater substructure. The concept also includes a novel design regarding the power export system based on long self-life and low-weight dynamic cables. The project also includes interesting features in the field of sensoring and monitoring of both structures and cables and the evaluation of the techno-economic, environmental, social, and socio-economic impacts associated with the development of a commercial wind farm using the FLOTANT concept.

The FLOTANT technology aims to achieve an overall 60% reduction in CAPEX and 55% in OPEX allowing a competitive LCoE in the range of 85-95 €/MWh by 2030 [2].

2. Sites

The FLOTANT platform has been designed for two different locations [3]. The first of them is considered to represent the ideal conditions for commissioning a floating concept based on a barge-type floater, i.e., an area with very constant winds and low extreme wind speeds. The selected location is off the southeast coast of Gran Canaria (GC) island (Figure 1a), in the Canary Islands, Spain. The exact location and the principal site characteristics are provided in Table 1.

The second site selected to study the behaviour of the FLOTANT concept is located 19 km West of Barra (WoB) Island (Figure 1b), Scotland. This site is considered a high energy environment and was identified by a previous project funded by the European Commission under the Horizon 2020 programme as a potential area for the development of deep-water floating technology. The principal characteristics and accurate site location are also provided in Table 1.

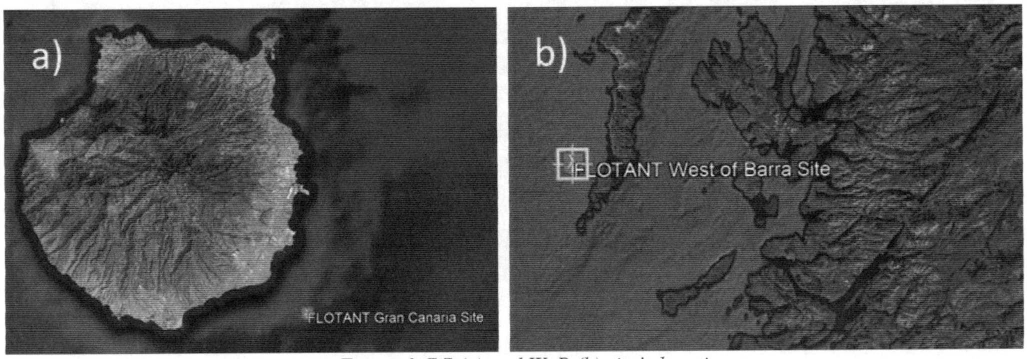

Figure 1 GC (a) and WoB (b) site's location

Table 1 GC and WoB sites characteristics

Item	Unit	GC	WoB
Water Depth	m	250.00	100.00
V_{50}	m/s	28.00	50.00
Hs_{50}	m	5.11	15.60
Tp_{50}	s	12.00	15.20
Seabed type	-	Sand	Basalt

3. Wind turbine

The wind turbine developed for the FLOTANT project is the INS12MW generic wind turbine (GWT) resulting from the upscaling exercise of the DTU10MW reference wind turbine (RWT), an existing public aeroelastic model designed by the DTU as part of the Light Rotor project [4]. Table 2 presents the main properties of INS12MW GWT [5]. Figure 2 depicts thrust and the steady-state behaviour of the INS12MW GWT.

Table 2 Principal wind turbine characteristics

Item	Unit	INS12MW GWT
Output power	MW	12.00
Rotor diameter	m	195.40
Hub diameter	m	6.13
Hub height above sea level	m	119.70
Tower mass	t	782.00
Nacelle mass including rotor (RNA)	t	836.00
Distance from Tower Top to Hub Height	m	3.01
Cut-in wind speed	m/s	4.00
Rated wind speed	m/s	11.40
Cut-out wind speed	m/s	25.00
Cut-in rotor speed	rpm	3.50
Rated rotor speed	rpm	8.76

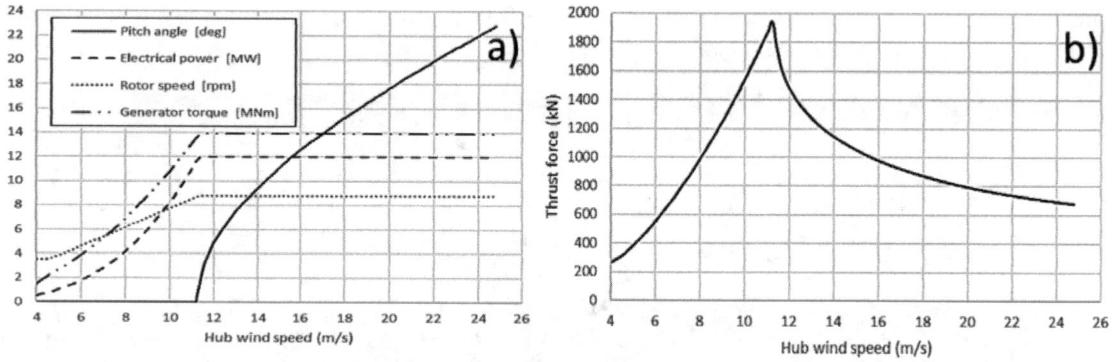

Figure 2 Main steady-state parameters (a) and thrust force (b) versus wind speed

3.1. Aeroelastic modelling

Using OpenFAST [6], the open-source glue code developed by NREL, an aeroelastic model of the INS12MW wind turbine has been developed. OpenFAST uses the results of several pre-processors (e.g., IECWind, TurbSim among others), and combines them within several simulations (e.g., ElastoDyn, InflowWind, AeroDyn, ServoDyn among others) to calculate the coupled dynamic response of the wind turbine.

The pre-processors are tools designed for helping to create aero-elastic models. They produce relevant information needed to feed the simulation tools. IECWind is a utility program used to create wind files that model the extreme conditions outlined in [7] for AeroDyn-based programs. TurbSim is a stochastic, full-field, turbulent-wind simulator using a statistical model to generate time series of three-component wind speed for AeroDyn-based codes such as OpenFAST.

As stated before, OpenFAST couples results from different simulations. ElastoDyn is a dynamic structural model able to represent the rotor, drivetrain, nacelle, tower, and platform. It computes displacement, velocities, accelerations, and reactions from the acting loads taking into account the controller and the substructure reactions. InflowWind is a module for processing wind-inflow data generated in IECWind or TurbSim pre-processors. AeroDyn is a time-domain module that computes aerodynamic loads of horizontal axis wind turbines, and ServoDyn is a control and electrical drive model for blade pitch, generator torque, nacelle yaw, highspeed shaft brake and blade tip brake. A control system based on the DTU10MW controller has been tuned for the FLOTANT concept using the NREL's Reference OpenSource Controller (ROSCO) toolbox for wind turbine applications.

3.2. Design Load Cases (DLCs)

To check if the floating concept fulfils the requirements (Table 3) agreed by the consortia, a relevant subset of DLCs covering the global performance of the floater and the moorings behaviour was selected. The DLCs considered are presented in Table 4.

4. Floater

The FLOTANT platform is a concrete-plastic hybrid barge-type concept with a passive anti-flooding system (Figure 3). The floater is not watertight, and the floatability is ensured by filling the empty spaces within the floater using a low long-term water absorption extruded polystyrene material with a density of 0.95t/m3 [11].

The substructure contains a central concrete column linked to the external walls through concrete braces. The void space in between the braces is filled with the XPS foam material which is fixed to the bottom concrete slab by strands. The concrete tower provides the structural connection between the steel tower of the wind turbine with the bottom concrete slab. Six water ballast deposits are also fitted in the inner floater structure to help to compensate for the wind thrust. A heave plate is attached at the bottom of the floater to help with the heave motions control. The main dimensions of the floater are provided in Table 5 and depicted in Figure 3a.

Table 3 Motion and acceleration design criteria

DoF	Max	Mean	Standard Deviation	Max in Survival Condition	Mean in Survival Condition	Acceleration at the nacelle
Surge	±30 m	-	-	-	-	2,94 m/s²
Sway	±30 m	-	-	-	-	
Heave	-	-	-	-	-	
Roll	±5°	±2,5°	1°	±7°	±5°	-
Pitch	±5°	±2,5°	1,5°	±7°	±5°	-
Yaw	±15°	-	-	-	-	-

Table 4 DLC list considered in the final loop of design

DLC	**Comments**	**Wind/Sea state/Current model**	**Wind Speed [m/s]**	**Yaw error [°]**	**Wind/ Wave mis-alignment [°]**	**Wave directions [°]**	**Number of simulations**
1.2	Turbulence + normal sea state	NTM/NSS/ NCM	4to25@4	0	30; 0	0to60 @30	36
1.6	Turbulence + severe sea state	NTM/SSS/ NCM	9,4 / 11,4 / 13,4 / 25	8; 0	30; 0	-30to90 @30	160
2.3	Extreme gust + grid loss	EOG/NSS/ NCM	9,4 / 11,4 / 13,4 / 25	8; 0	0	30	24
2.4	Turbulence + grid loss	NTM/NSS/ NCM	4to25@4	0	0	30	6
2.6	One mooring line damaged	NTM/NSS/ NCM	9,4 / 11,4 / 13,4	-8	-30	30	18
2.8	Supply vessel impact, 2 ballast tanks flooded	NTM/NSS/ NCM	9,4 / 11,4 / 13,4	-8	-30	30	18
4.1	Normal wind conditions + normal sea state	NWP/NSS /NCM	4to25@4	0	0	30	18
6.1	Extreme wind and sea state	EWM/ESS /ECM	V50	8; 0	30; 0	-30to90 @30	40
6.3	Extreme wind and sea state + yaw misalignment	EWM/ESS /ECM	V1	20	30; 0	0to60 @30	36
6.4	Turbulence	NTM/NSS /NCM	1 / 3 / Va / Vb	0	30; 0	0to60 @30	24
7.1	Extreme wind and sea state + grid loss	EWM/ESS /ECM	V1	8; 0	30; 0	0to60 @30	24
7.3	One mooring line damaged	EWM/ESS /ECM	V1	-8	-30	30	6
7.5	Supply vessel impact, 2 ballast tanks flooded	EWM/ESS /ECM	V1	-8	-30	30	6

Table 5 Dimensions of the floater

Characteristics	Units	Values
Floater main diameter (D1)	m	48
Heave plate diameter (D2)	m	55.2
Concrete tower diameter (D3)	m	11.5
Concrete slab and heave plate height (H1)	m	2
Floater height (H2)	m	18
Concrete tower base height (H3)	m	16
Concrete tower top height (H4)	m	9
Draft	m	12
Displaced Volume	m^3	22028.85
Mass	t	21403.70

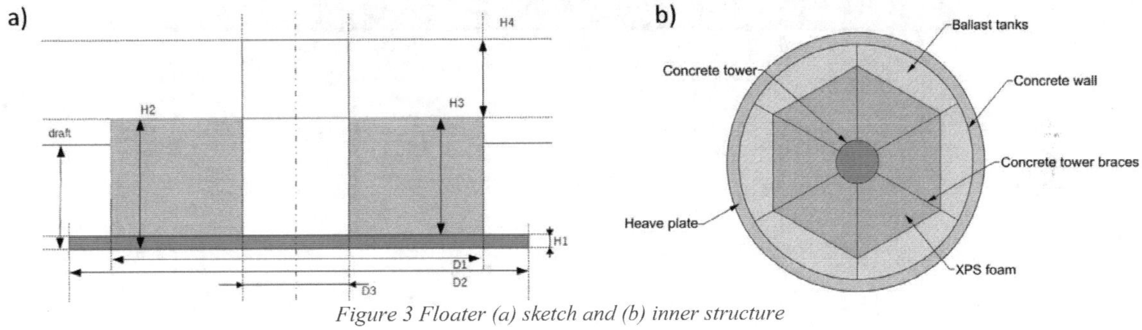

Figure 3 Floater (a) sketch and (b) inner structure

4.1. *Ballast system*

The ballast system features both active and passive water ballast. It comprises a set of six steel ballast tanks (1,218.86m^3 each) filled with water (Figure 3b).

Part of the ballast water is considered as passive ballast (728 t) and therefore, it remains in its position regardless of the environmental conditions whereas another part of the ballast water, the active ballast (257.8 t), can be moved around the different ballast tanks to counteract the mean pitch and roll induced by the thrust of the wind turbine. Such a system can correct up to 11.8 degrees of mean rotation even though only 8.5 degrees are required. Therefore, further optimization of the system will be required.

4.2. *Hydrodynamics and mooring lines*

In this research, the FASTLink module is used to couple OpenFAST and OrcaFlex. When the FAST-OrcaFlex Interface [8] is used, all hydrodynamic and mooring loads are computed by OrcaFlex [9]. To use OrcaFlex, a hydrodynamic database (HDB) must be loaded in advance. OrcaWave is the tool selected to work out the HDBs employed during the project. Note that several different HDB must be set for the different conditions under analysis.

The mooring system designed for GC consists of four semi-taut lines using wire rope whereas for WoB a chain catenary system with five lines was developed (Table 6). The top view of the WoB setup is similar to the GC one (Figure 4) but presents three front lines instead of two, to be able to cope with the harsher environmental loads. All the lines in both locations are equipped with the innovative elastomer springs [11] shown in red in Figure 4. All mooring lines are attached to the bottom outside edge of the FLOTANT platform and fixed to the seabed by drag anchors in GC and gravity base caisson anchors in WoB. Fairlead and anchor positions are provided in Table 7.

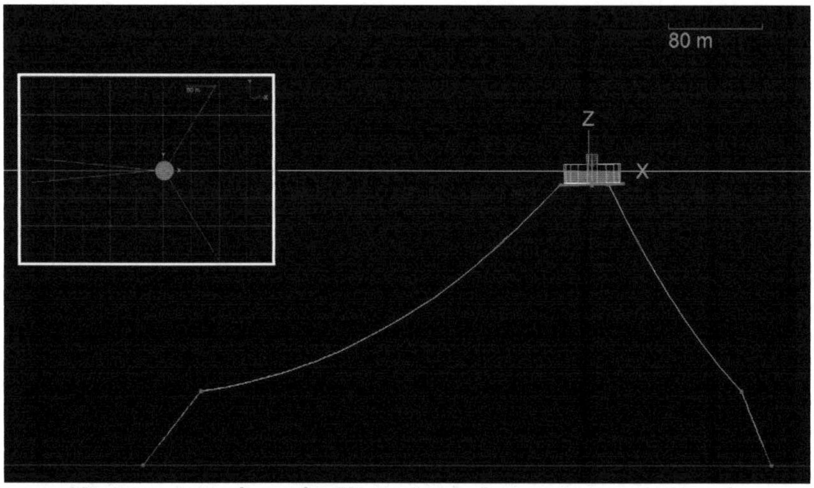

Figure 4 mooring layout for GC site. Spring elastomers are marked in red.

Table 6 Mooring system parameters

Characteristic	Unit	GC	WoB
Wire (GC) and chain (WoB) outer diameter	mm	91.12	0.27
Mass density in air	kg/m	51.85	430
Axial stiffness	MN	525.00	1845
Seabed friction coefficient	-	0.50	0.50
Normal added mass coefficient	-	1.00	1.00
Normal drag coefficient	-	1.60	2.40
Axial drag coefficient	-	0.008	1.15
Unstretched length of frontlines	m	438.00	880.00/855.00/ 880.00
Unstretched length of backlines	m	361.14	952.00
Elastomer length	m	16	20

Table 7 Fairlead and anchor positions in meters from the centre of the floater and waterline

Site	Line	Fairlead x	Fairlead y	Fairlead z	Anchor x	Anchor y	Anchor z
GC	1	-27.6	0.00	-10.99	-383.18	-34.27	-250.00
	2	-27.6	00.00	-10.99	-383.18	34.27	-250.00
	3	13.8	-23.90	-10.99	155.94	-255.34	-250.00
	4	13.8	23.90	-10.99	155.94	255.34	-250.00
WoB	1	-27.6	00.00	-10.99	-847.01	0.00	-100.00
	2	-27.6	00.00	-10.99	-839.26	-233.90	-100.00
	3	-27.6	0.00	-10.99	-839.26	233.90	-100.00
	4	15.83	-22.61	-10.99	604.00	-738.76	-100.00
	5	15.83	22.61	-10.99	604.00	738.76	-100.00

4.3. *Hydrodynamics: Panel pressures*

The time-domain hydrodynamic pressure at the floater panels has been calculated since it is required for subsequent structural calculations. The hydrodynamic pressure calculated considers the sum of:

- Hydrostatic pressure around the hull
- Incident pressure from incoming waves
- Diffraction pressure due to interaction between waves and hull
- Radiation pressure due to waves created by the hull motion

4.4. Stability

This section presents the method and a summary of the results of the stability analysis. The analysis has been performed using the commercial software General HydroStatics (GHS). GHS is a code able to evaluate the flotation, trim, and stability of floating structures.

The stability analysis aims to assess the safety and performance of the FLOTANT concept against the design criteria detailed in [12] and considers the actions of static wind and hydrostatic forces in different system conditions. The study encompasses the assessment of the stability of the FLOTANT concept across the Pre-Service (construction), In-Service (production, survival, and open sea maintenance), and Non-Service conditions (transportation, installation, maintenance in protected sites and commissioning).

The FLOTANT concept fulfils the design criteria abovementioned in all the conditions studied. The In-Service conditions at WoB are the most challenging ones. Figure 5 depicts the righting arm and heeling arm curve graph for WoB considering 11.4m/s wind speed and headwind direction with ballast split equally between all tanks. This is the most challenging situation considered in the present analysis and comfortably passes the stability requirements demanding the area under the righting moment curve to the second intercept is equal to or greater than 140% of the area under the wind heeling moment curve. Subsequently, different analyses were performed considering only the required tank filled with the active ballast and different yaw misalignment angles. The platform fulfils stability requirements within a range of misalignment between -65° and 65° approximately (relative to the wind direction).

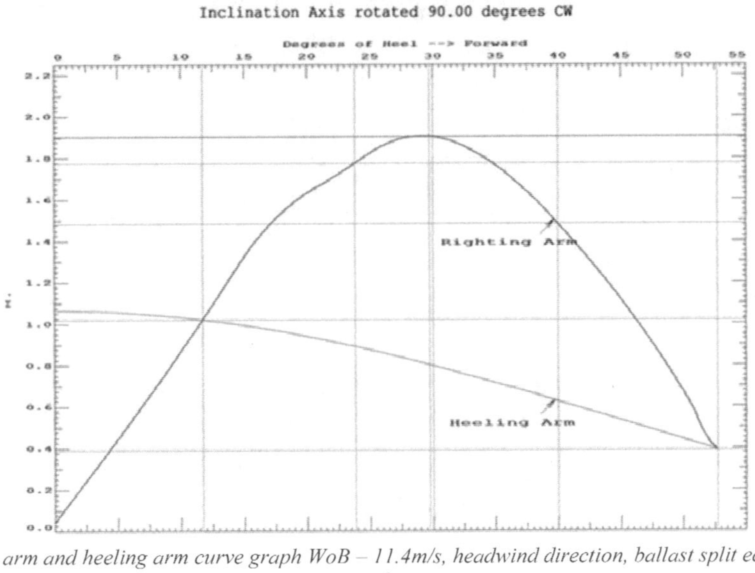

Figure 5 Righting arm and heeling arm curve graph WoB – 11.4m/s, headwind direction, ballast split equally between all tanks

Besides, the maximum Vertical Centre of Gravity (VCG) curves have been calculated showing the maximum possible overall VCG of the FLOTANT platform for a reasonable range of displacements shedding acceptable values.

4.5. Fatigue Limit State calculation

Fatigue resistance has been checked in two sections of the concept: the interface between the steel and concrete tower and the section of the concrete tower above the concrete tower braces. To calculate the Fatigue Limit State (FLS) of the abovementioned concrete elements, Markov matrices have been employed since the fatigue calculations for concrete require both, the mean stress value and the cycle stress range (local maximum and minimum) for each load cycle.

Markov matrices have been computed following the requirements stated in [13] considering DLCs 1.2, 2.4, 4.1, and 6.4 presented in Table 4 above.

5. Global performance results

This section presents a summary of the most relevant results from the analysis of the simulations of the integrated model of the FLOTANT concept.

5.1. Summary results

Table 8 presents a summary of the cases fulfilling the requirements presented in Table 3 considering the results of the simulations listed in Table 4 for the two locations under study.

Table 8 Percentage of cases satisfying the abovementioned requirements

	Surge (Max)	Sway (Max)	Roll (Max)	Roll (AVG)	Roll (SD)	Pitch (Max)	Pitch (AVG)	Pith (SD)	Yaw (Max)	Acc x (Max)	Acc y (Max)
%GC	100	100	94	98	60	95	98	92	100	100	100
%WoB	100	80	91	96	42	95	96	83	100	95	100

5.2. Load matrix

Table 9 presents the diagonals of the ultimate limit state load matrix obtained from the coupled simulations for GC and WoB sites for the tower base. The load matrix has been computed as required in [13].

Table 9 ULS load matrix

	Fx (kN)	Fy (kN)	Fz (kN)	Fxy (kN)	Mx (kN.m)	My (kN.m)	Mz (kN.m)	Mxy (kN.m)
Max GC	7746	7942	-17132	8096	410670	650160	33318	688743
Min GC	-4123	-4852	-25596	804	-663120	-335340	-41499	28031
Max WoB	10391	11553	-11872	11556	651510	835920	45698	777455
Min WoB	-9060	-8073	-30686	2024	-919620	-733050	-45144	194402

5.3. Panel pressures

Figure 6 below depicts the result of the panel pressures calculated for an instant of time within DLC 1.6.

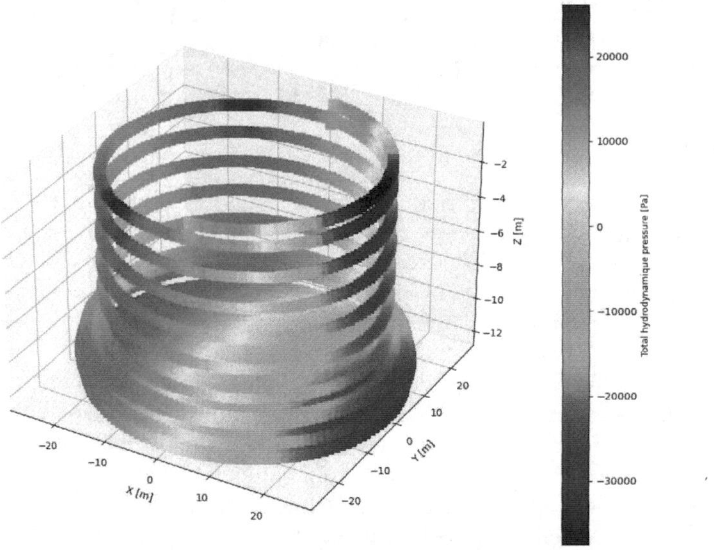

Figure 6 Panel pressures calculated for DLC 1.6.

5.4. Fatigue loads: Markov matrix

Figure 7 below depicts a representation of the Markov matrix calculated for DLC 1.2.

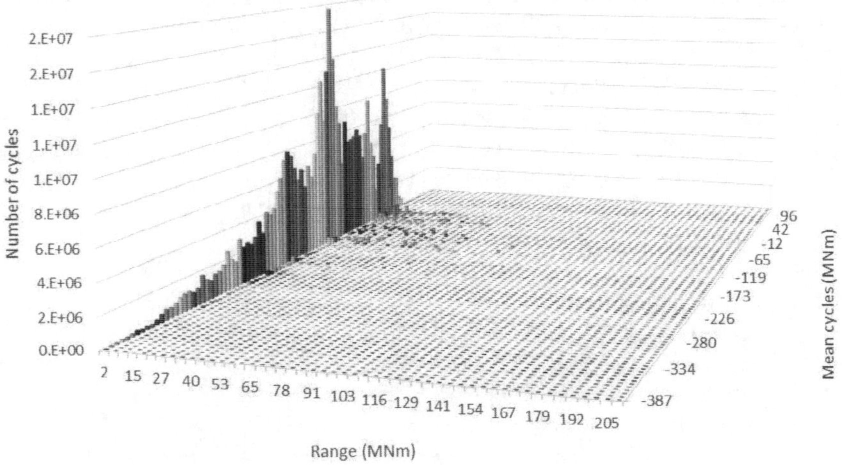

Figure 7 Markov matrix for DLC1.2.

6. Conclusions and recommendations

The INS12MW GWT has been set up and presented along with the main characteristics of the FLOTANT floater and the mooring system optimized for deep waters.

The FLOTANT concept behaves satisfactorily from a general point of view presenting both low motions in the floater and low accelerations in the nacelle. The percentage of DLCs passing the restrictive criteria imposed by the FLOTANT consortia is adequate to the development stage of the concept. There is a remarkable trend increasing the percentage of cases fulfilling the criteria when compared with the first iteration design loop [1]. However, as presented in the summary of results, a few DLCs do not meet the criteria and therefore more in-deep analysis is required:

- The main reason behind many of the cases not fulfilling the requirements is the lack of requirements definition for certain conditions which have been encompassed within 'Survival' and required specific requirement definition e.g., line loss and supply vessel impact. These DLCs should have specific requirements since survival requirements are aligned to the desired response under extreme conditions rather than the desired response in accidental conditions.
- Beyond that, the mooring design is not stiff enough in the sway direction. A less directional mooring layout is suggested for the next optimization loop.
- Especially in the GC site, the dynamic behaviour is slightly poorer than the one presented in the first iteration design loop [1]. Therefore, it should be concluded that the previous catenary system suits better the FLOTANT concept than the current semi-taut configuration.
- Including buoys and weights into the mooring layout will help to achieve rotational motion improvements.

An example of load matrix, instantaneous panel pressures calculation result, and a Markov matrix worked out with the results of the DLC1.2 have been presented to show the kind of results needed to perform the subsequent structural analysis [3].

Acknowledgements

This publication has been supported by the H2020 project FLOTANT coordinated by the Oceanic Platform of the Canary Islands (PLOCAN). The project has received funding from the European Union's Horizon 2020 research and innovation programme under grant agreement No. 815289

References

[1] Serret J, Kahn B, Cavanagh B, Lorente P, Pascal R, Girandier C, Cortes C, Duran R, McEvoy P and Castro A 2021. First Iteration Design of the Flotant Concept. *Proceedings of the ASME 2021 40th International Conference on Ocean, Offshore and Arctic Engineering*, OMAE2021
[2] FLOTANT 2019. *Innovative, Low Cost, Low Weight And Safe Floating Wind Technology Optimized For Deep Water Wind Sites.*
[3] FLOTANT 2020. *D4.1 – Structural And Naval Architecture Design Basis.*
[4] Bak C, Zahle F, Bitsche R, Kim T, Yde A, Henriksen L C, Hansen M.H, Blasques J, Gaunaa M and Natarajan A 2013. *The DTU 10-MW Reference Wind Turbine.*
[5] FLOTANT 2019. D4.2 - *Design Brief: Specifications of a generic wind turbine.*
[6] Jonkman J and Buhl M 2005. FAST user's guide. Golden, Co.
[7] IEC 61400-3 2009. *Wind turbines - Design requirements for offshore wind turbines*. Geneva.
[8] NREL NWTC Information Portal (FAST-OrcaFlex Interface) 2011. Golden, Co.
[9] ORCINA n.d. *OrcaFlex for offshore wind.* Technical Notes. Ulverston, UK.
[10] FLOTANT 2020. *D2.1 – Mooring and Anchoring System Design.*
[11] FLOTANT 2019. *D4.4 – Floater Predesign*
[12] DNVGL-ST-0119 2018. *Floating wind turbine structures*. Oslo.
[13] DNVGL-ST-0437 2016. *Loads and site conditions for wind turbines*. Oslo.

Predictive Maintenance of Floating Offshore Wind Turbine Mooring Lines using Deep Neural Networks

N Gorostidi[1,2], V Nava[1,3], A Aristondo[3] and D Pardo[2,1,4]

[1] Basque Center for Applied Mathematics, Alameda Mazarredo 14, 48009 Bilbao, Spain
[2] University of the Basque Country (UPV/EHU), Barrio Sarriena S/N, 48940 Leioa, Spain
[3] TECNALIA, Basque Research and Technology Alliance (BRTA), Astondo Bidea, Edificio 700, Derio, 48160, Spain
[4] Ikerbasque (Basque Foundation for Sciences), Bilbao, Spain

E-mail: ngorostidi@bcamath.org

Abstract. The recent massive deployment of onshore wind farms has caused controversy to arise mainly around the issues of land occupation, noise and visual pollution and impact on wildlife. Fixed offshore turbines, albeit beneficial in those aspects, become economically unfeasible when installed far away from coastlines. The possibility of installing floating offshore wind turbines is currently hindered by their excessive operation and maintenance costs. We have developed a comprehensive model to help companies plan their operations in advance by detecting failure in mooring lines in almost real time using supervised deep learning techniques. Given the lack of real data, we have coupled numerical methods and OpenFAST simulations to build a dataset containing the displacements and rotations of a turbine's floating platform across all directions. These time series and their corresponding frequency spectra are used to obtain a set of key statistical parameters, including means and standard deviations, peak frequencies, and several relevant momenta. We have designed and trained a Deep Neural Network to understand and distinguish amongst a series of common failure modes for mooring lines considering a range of metocean and structural conditions. We have obtained promising results when monitoring severe changes in the line's mass and damping using short time spans, achieving a 95.7% validation accuracy when detecting severe biofouling failure.

1. Introduction

The rapid development of cost-effective, green forms of energy is one of humanity's main assets in the fight for environmental conservation. In particular, wind power systems play a huge role in the current scenario. Controversy, however, has recently arisen following the worldwide massive installation of onshore wind turbines, mainly around issues such as land occupation, impact on wildlife and noise and visual pollution. That is one of the many reasons for which research has turned its focus towards offshore wind rather than onshore wind. Fixed offshore turbines are already being deployed all around the world but, despite their environmental benefits, their installation costs increase exponentially when moving further away into the seas, as seabed depth escalates. Floating Offshore Wind Turbines (FOWT), albeit more affordable in deeper waters than their fixed counterparts, are still economically unfeasible due to their excessive Operation and Maintenance (O&M) costs. In fact, Figure 1 suggests O&M accounts for approximately a third of the total costs associated to a floating wind project [1].

Figure 1. Simplified cost breakdown of a floating offshore wind project [1].

Cost optimisation is one of the main reasons for which Structural Health Monitoring (SHM) techniques are used in industry. This paper applies the notion of spectral analysis, whose first implementation for SHM purposes dates back to the 1960s, when Lifshitz and Rotem [2] observed dynamic modulus variations under vibrating loads to detect damage in composite materials. This technique has ever since then been implemented for an extensive range of engineering disciplines, including amongst others the aerospace and civil industries [3]. This condition-based monitoring approach has been widely applied to floating oil platforms [4]. In fact, Prislin and Maroju [5] implemented a machine learning model to evaluate the integrity of mooring systems for oil platforms. Over the past few years, researchers have focused on the concept of Failure Modes and Effects Analysis (FMEA) to establish strategic criteria for maintenance operations by relating the frequency of different forms of damage with their corresponding economic impact [6, 7].

The world is currently turning its eyes towards big data and artificial intelligence, and wind power research is no exception. As the ever-increasing computational resources allow for previously inconceivable calculations, researchers are nowadays implementing deep learning algorithms to detect damages in different components for wind systems [8]. Coupling deep neural networks and vibration-based fundamentals for the structural health monitoring of floating wind turbine mooring lines nonetheless constitutes a complete innovation in the field. Our research applies some of the concepts introduced by Li et al. [9], who presented a model to recreate the response of a floating platform under different mooring configurations; and Jaiswal and Ruskin [10], who implemented computer vision techniques to detect floating vessel mooring failure. Martinez-Luengo et al. [11] and Joshuva et al. [12] have extensively reviewed other methods applied for O&M cost minimisation purposes, including for instance thermal imaging and acoustic emission monitoring.

This research aims at extending the work presented by Gorostidi and Nava [13] by developing an AI-based algorithm to predict failure in FOWT mooring lines. From the displacements and rotations of the turbine's platform, a series of key modal parameters is identified and fed into a deep learning structure [14], which is ultimately able to discern whether a mooring line is failing or not, and what kind of damage is affecting its performance. This approach is intended to help companies plan their operations in advance, and could potentially reduce their total costs substantially by keeping unnecessary commissions and sensorization costs to a minimum.

The main limitations of the method we use arise from the fact that floating offshore wind turbines are still not mass-produced as of today, and thus real data is very limited. Instead, we carry out numerical simulations to recreate the external and structural characteristics affecting a turbine's behaviour as truthfully as possible, but the model's performance under real conditions still remains a challenge. Furthermore, we have built a neural network using a classification approach. As with any supervised learning technique, this method relies on establishing labels for the potential forms and severities of damage. The main advantage of these models with respect to unsupervised algorithms is the fact that we are able to determine what kind of damage is affecting a turbine's performance rather than just detecting anomalies. However, classifying failure implies imposing a binary choice of whether a mooring system is damaged or not. This poses the challenge of setting a discrete threshold for the minimum damage severity at which repairs should be performed. A set of intermediate labels should therefore be simulated and coupled with a compatible reliability model so as to estimate at what point in time a given mooring system will transition from an intermediate state to a severely damaged one.

2. Methodology

We divide the method's workflow into two phases: simulation and training. The former includes all preprocessing, simulation and data postprocessing operations that end once the training and validation datasets are generated. This step is necessary given the current scarcity of real data. In the training phase we design and train a Deep Neural Network using said datasets to build an effective algorithm for the prediction of mooring line damage. A short overview of the initial stages of the project is given to introduce the studied scenario and some of the model's key variables.

2.1. Preliminary approach

The introductory work published in [13] presented a simple 1-DOF differential equation, which we used to recreate the dynamics of a generic FOWT platform.

$$M \cdot \ddot{x} + C \cdot \dot{x} + K_3 \cdot x^3 + K_1 \cdot x + K_0 = F_{wave}\left(H_S, T_P, t\right) + F_{drift}\left(H_S^2\right) + F_{wind}\left(V\right). \quad (1)$$

In this instance we only studied the platform's surge. A set of ranging metocean parameters, including wind velocity and the waves' significant height and peak period, was defined to create an extensive set of training data. This contained a collection of synthetic statistics extracted from the platform's response in both time and frequency domains. A series of alterations to the system's original structural characteristics was defined to simulate a range of failing configurations. These included biofouling damage to the mooring lines, resulting in increased mass and damping coefficients, which eventually yielded longer periods and hence lower peak frequencies; pitting wear, and anchoring point displacements, which caused different effects on the line's non-linear stiffness coefficients.

2.2. Simulation and dataset generation

Our research currently extends this idea by considering not only the longitudinal displacements of the floating platform, but a fully comprehensive 6-DOF system in which all transverse effects and rotations are included. We therefore consider six time series, from which we calculate and extract statistics such as means and standard deviations, peak frequencies and some relevant momenta. These parameters are ultimately the inputs to the implemented deep learning algorithm, which then infers whether the mooring lines are damaged or not, and if so, what causes might be affecting their performance. Figure 2 presents a simplified diagram of the method's workflow.

We have generated the training and validation datasets by using NREL's open-source wind turbine simulation tool OpenFAST [15] to perform simulations on the semisubmersible floating system depicted in Figure 3(a), developed within the DeepCwind project [16]. Considering a simple catenary mooring system, we computed a series of scenarios combining both metocean and structural parameters by rewriting the modules shown in Figure 3(b). In particular, *InflowWind* was modified to adjust the horizontal, steady wind velocity. *HydroDyn* was used to generate the

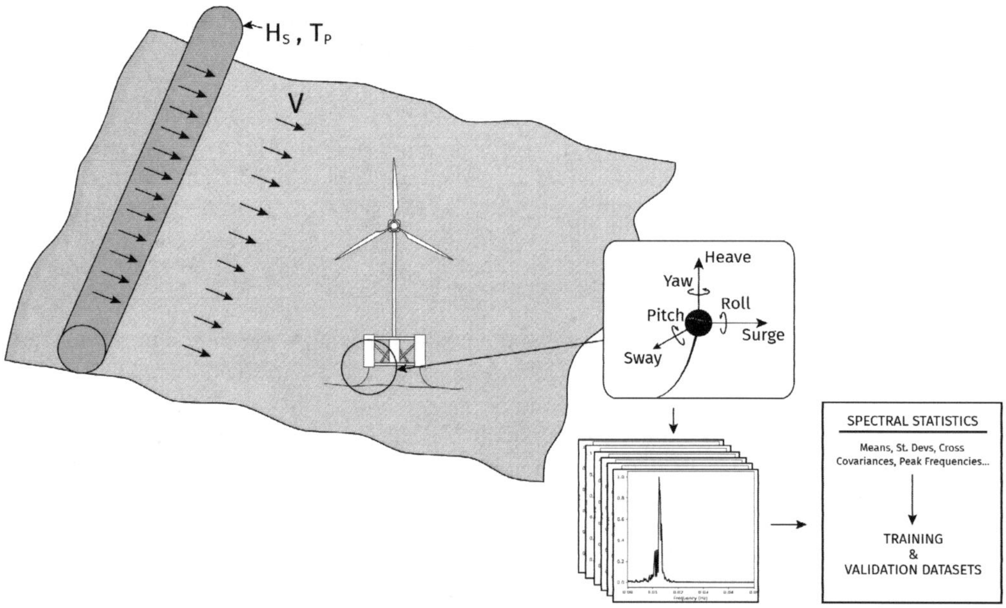

Figure 2. Schematic representation of the method's workflow, including simulation and data extraction and visualisation processes.

time domain hydrodynamics behaviour of the platform in different sea states, characterised by the significant wave height, H_S, and peak period, T_P. The structural properties of the mooring lines, e.g. mass, damping and stiffness, were modified using *MoorDyn*.

Thanks to the *ElastoDyn* module, we simulated the time evolution of the platform's displacements across all six degrees of freedom. Their respective frequency spectra were calculated using a postprocessing script. From these, the desired datasets were built, containing the statistics shown in Table 1. The implemented network is designed to be agnostic to external factors. This means that wind and wave characteristics are dropped out of the dataset before training, and are only included at first for data analysis purposes, e.g. filtering and visualisation. The result of this is an algorithm capable of predicting FOWT mooring failure without needing any sensors to describe wind and wave conditions, thus saving even more on measuring costs. The model is therefore trained using mean, standard deviation, peak frequency and zero-order momentum of the response for each of the 6 DOFs, hence resulting in 24 total dataset features.

In the end, the aim is for the DNN to identify discrepancies in these modal parameters for different mooring line health states. To give an oversimplistic example, an unexpectedly heavy line might cause inertial effects to increase, thus causing the platform to move more slowly than predicted. This movement is then characterised by longer periods, and hence smaller frequencies. Once trained, the implemented network should therefore understand and associate lower peak frequencies to this kind of damage, which could possibly arise from biofouling issues.

2.3. Network design and training

We use supervised deep learning techniques to model this problem by imposing a series of changes to the mooring line's structural properties. More specifically, we employ a simple feed-forward, fully-connected deep neural network following the topology shown in Figure 4. As stated before, its inputs, denoted by i_m, are the modal statistics obtained from the floating

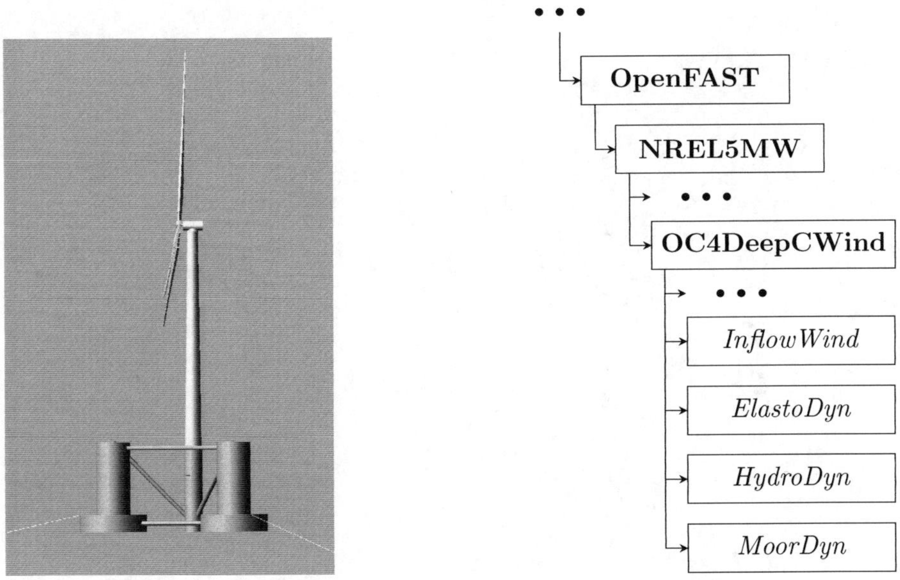

(a) (b)

Figure 3. Simulation setup in OpenFAST (a) Sketch of the DeepCwind semisubmersible floating platform (b) OpenFAST's default folder structure.

platform's displacements. The outputs, indicated as p_k, are the supplementary probabilities of the given line being assigned to any of the k labels or, in this case, health status scenarios. The idea, once the network has been trained, is for one of these probabilities to be as close as possible to one, with the rest being approximately equal to zero.

The algorithm designed in [13] contained all the damaged configurations shown in Table 2. The implemented 6-DOF model considers fully-functioning mooring lines, as well as those affected by biofouling issues, that is, with increased mass due to the undesired attachment of

Table 1. Parameters stored in the training and validation datasets. Significant wave height ranged from 4 to 10 m. Peak wave period, from 5 to 15 s. Wind speed varied from 2 to 15 m/s.

Symbol	Parameter
$\bar{x}_r$	Response mean
σ_r	Response standard deviation
f_P	Peak frequency
m_0	Zero-order momentum
H_S	Significant wave height
T_P	Peak period
V	Wind speed
Label	Health status label

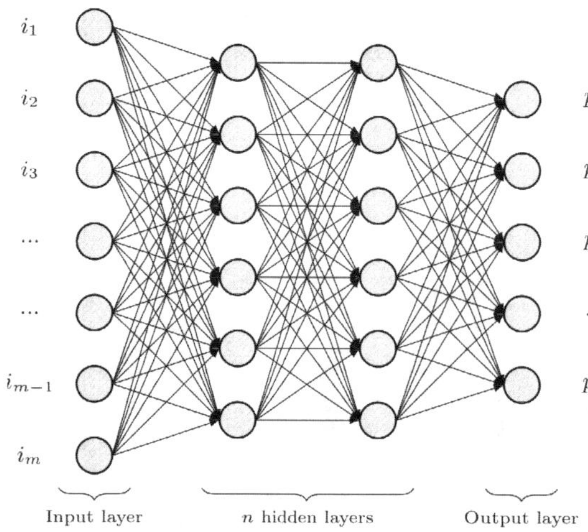

Figure 4. Generic topology of a four-layer, feed-forward deep neural network.

marine life. This condition is recreated in OpenFAST by specifying an increment in mass per unit length in *MoorDyn*, whose default value is 113.35 kg/m. In this instance, we have considered a 10% mass increase to constitute severe biofouling failure. The damaged lines therefore have a mass per unit length of 124.69 kg/m, with all the consequences this causes in the dynamics of the mooring system.

A total of 3,140 cases were computed from OpenFAST simulations combining different wave and wind conditions for both undamaged and damaged lines. The obtained signals have a duration of three hours to mitigate any transient and/or unsteady effects, following the conclusions obtained in [13]. A dataset containing 2,355 samples, that is, 75% of the total computed cases, was built to train the network, while the remaining 785 samples were used to simultaneously validate the model. These validation cases do not affect neither the algorithm's training process nor its coefficients, and are only employed to estimate the performance of the model with respect to unseen data. The implemented neural network was tuned using the hyperparameter values shown in Table 3, all of which are standard for classification tasks with only two classes.

Table 2. Simulated damage configurations for the health status of mooring systems for both the initial and current models.

Condition	Definition	1-DOF	6-DOF
Anchoring	Displaced mooring line anchor	✓	
Biofouling	Attached mussels, algae and other marine organisms	✓	✓
Fatigue	Damaged links in a mooring line due to wear or fatigue	✓	
Undamaged	Default mooring line properties	✓	✓

Table 3. Hyperameters used in the training stage of the project.

Parameter	Definition
Neurons per layer	24, 16, 12, 2
Activation functions	ReLU, Softmax
Layer connection	Fully-connected layers
Optimiser and learning rate	Adam, 0.0001
Cost function	Binary cross-entropy
Early-stop criterion and patience	Validation loss, 500 epochs
Training epochs	10,000

3. Analysis of Results

We have carried out a series of OpenFAST simulations combining both external conditions and structural characteristics of a FOWT's mooring system. The displacements and rotations of the floating platform have been computed for each case, yielding a set of six time series such as the one plotted in Figure 5. The unsteady appearance of the signals is caused by the irregular nature of the waves, which have been computed using a Pierson-Moskowitz spectrum with second-order mean-drift forces. A predominantly sinusoidal behaviour can be observed for each individual degree of freedom, thus justifying the use of peak frequencies as a key statistic to include in the training process. Moreover, it is a positive sign that the patterns presented in all directions and rotations are consistent with other studies [17].

Looking at the specific values for the displacements of the floating platform, and given the one-dimensional constraint imposed for both waves and wind, it bodes well that the surge, shown in Figure 5(a), presents much higher mean and amplitude in comparison to the transverse movements, sway and heave, displayed in Figure 5(b) and Figure 5(c), respectively.

The previously implemented 1-DOF model was trained obtaining a promising 96% validation accuracy when severe biofouling and anchor displacement issues were considered [13]. Separate analyses were carried out on each failure mode individually to estimate the network's sensitivity to increasingly severe damage. The most relevant outcome from these studies is the fact that an approximately 92% accuracy was obtained after a binary classification test between default mooring lines and 10% heavier ones.

The current status of our research already considers the displacements and rotations of the turbine's floating platform across all directions, thus producing much more realistic results. The evolution of the algorithm's training and validation processes is shown in Figure 6(a) and Figure 6(b) for loss and accuracy, respectively. A plateau is reached after training the network for more than 6,000 epochs, yielding a validation accuracy of 95.7%. This is already a significant improvement from the previously computed results in [13], and is promising considering the increasing complexity of the data. These outcomes are numerically displayed on Table 4, which presents a series of standard metrics for classification problems. Precision is a way to measure the quality of the implemented model by computing the proportion of all the instances identified as *positive*, that is, damaged, that are appropriately defined. Recall measures the quantity of *positive* cases that the model is able to properly identify. F-Score is a weighted average of precision and recall, with F1-Score in particular being equal to their harmonic mean. A comprehensive analysis of the performance of any classification model discusses all of these metrics, bringing increased attention to either one depending on the nature of the problem. Recall is more sensitive with respect to false negatives, that is, failing mooring lines being wrongly identified as undamaged. That is the reason for which we should maximise this over

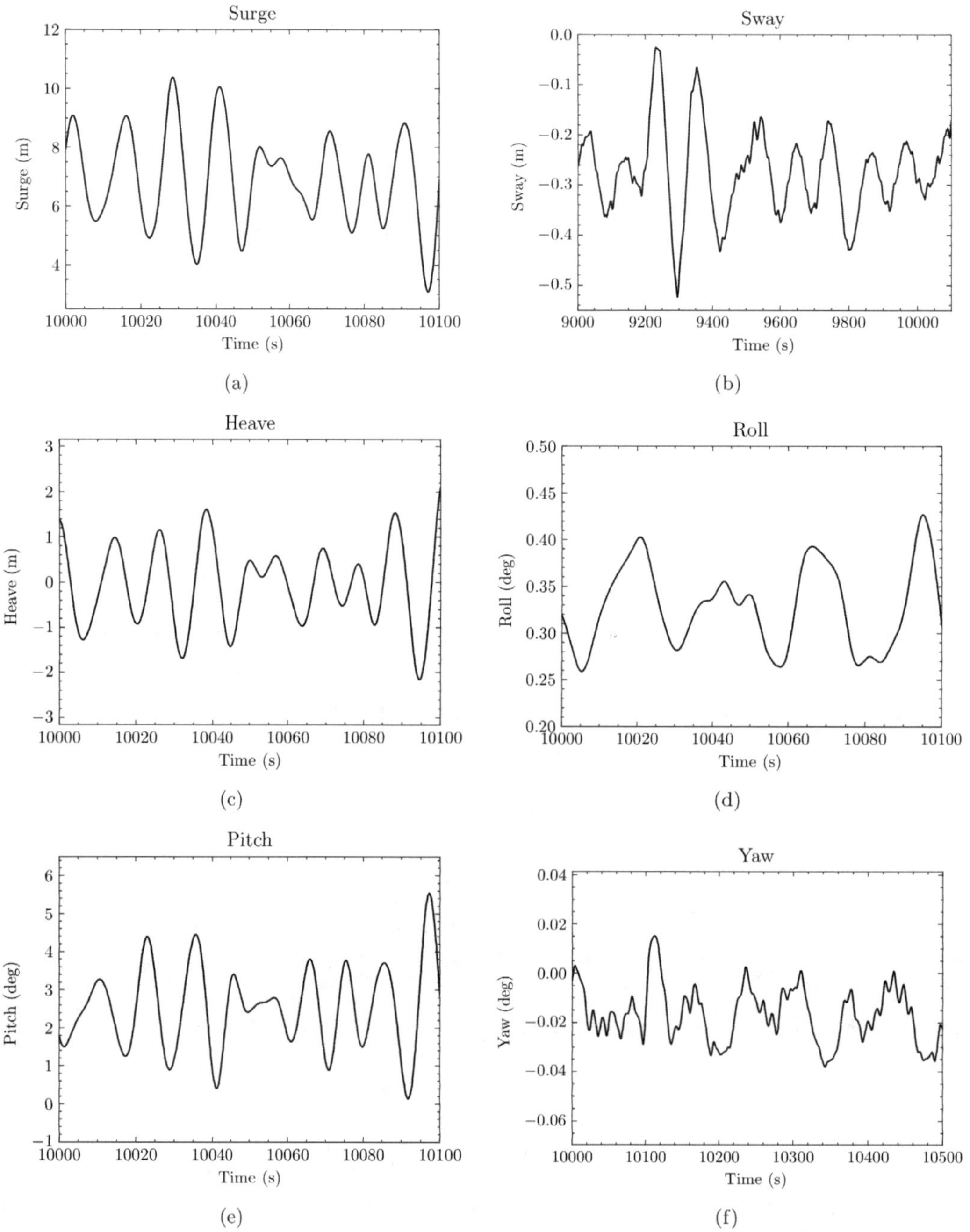

Figure 5. Observed displacements and rotations of a floating platform with undamaged mooring lines under $H_S = 10$ m, $T_P = 13$ s and $V = 8$ m/s: (a) surge, (b) sway, (c) heave, (d) roll, (e) pitch and (f) yaw. Sway and yaw have been plotted using a broader window for clarity.

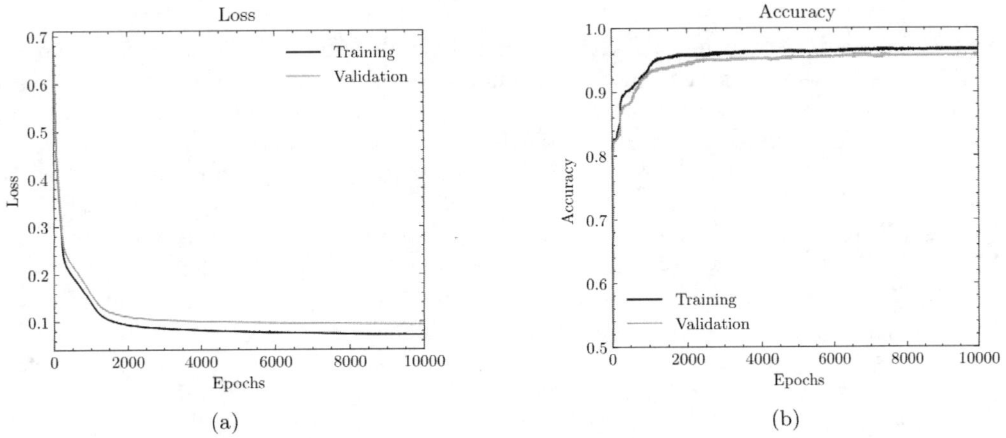

Figure 6. Evolution of training and validation (a) loss and (b) accuracy as training progresses.

any other metric in our scenario, as low recall would imply underestimating mooring line damage, thus hindering the model's potential benefits. The fact that the obtained 97.5% biofouling recall is the maximum out of all the metrics presented in Table 4 is very positive.

Another common way to illustrate the behaviour of a machine learning model is through the use of confusion matrices, such as the one shown on Table 5, which presents the algorithm's performance in a more intuitive manner. It shows that out of 400 damaged mooring systems provided by the validation dataset, only 10 have been classified as undamaged. The idea is for the model to eventually discern among a larger array of labels, thus being able to identify more kinds of damages.

Table 4. Model performance metrics for the validation set after 10,000 training epochs.

	Precision	Recall	F1-Score
Undamaged	0.973	0.938	0.955
Biofouling	0.942	0.975	0.958
Accuracy	0.957	0.957	0.957

Table 5. Confusion matrix for the validation set after 10,000 training epochs.

	Undamaged	Biofouling
Undamaged	361	24
Biofouling	10	390

4. Conclusions

We have developed a deep learning algorithm to predict failure in FOWT mooring lines using simple measurements. A series of numerical methods and OpenFAST simulations have been used to generate a database containing modal statistics from short-term displacements and rotations of a floating turbine's platform. We have defined a range of external conditions and structural characteristics to simulate different damaged mooring line configurations. We have designed and trained a deep neural network to so far detect biofouling failure in almost real time. The computed displacements show consistency and match those obtained in similar studies. The implemented network reaches over 95% accuracy when predicting moderate-to-high biofouling damage to a turbine's mooring lines. This spectral analysis could reduce O&M costs massively, thus increasing overall profitability, as only a few key sensors would be needed to remotely estimate the future of a mooring system's health status. The implemented method could potentially be extended to other subsystems of a floating offshore wind turbine, such as its blades, gearbox or tower.

Acknowledgments

N Gorostidi has received funding from the Spanish Ministry of Science and Innovation project DEEPINVERSE, with reference PID2019-108111RB-I00 (FEDER/AEI). V Nava has received funding from the project IA4TES - *Inteligencia Artificial para la Transición Energética Sostenible* funded by Ministry of Economic Affairs and Digital Transformation (MIA.2021.M04.0008); the "BCAM Severo Ochoa" accreditation of excellence (SEV-2017-0718); and the Basque Government through the BERC 2022-2025 program, the Elkartek project EXPERTIA (KK-2021/00048).

References

[1] Nava V, Ruiz-Minguela P, Perez-Moran G, Rodriguez Arias R, Lopez Mendia J and Villate-Martinez J L 2019 *Installation, operation and maintenance of offshore renewables* (Institution of Engineering and Technology) pp 397–424
[2] Lifshitz J and Rotem A 1969 *Journal of Composite Materials* **3** 412–423
[3] Doebling S, Farrar C and Prime M 1998 *The Shock and Vibration Digest* **30** 91–105
[4] Chang P, Flatau A and Liu S 2003 *Structural Health Monitoring - an International Journal* **2** 257–267
[5] Prislin I and Maroju S 2017 *OTC Offshore Technology Conference* **3** 1–12
[6] Carroll J, McDonald A and McMillan D 2016 *Wind Energy* **19** 1107–1119
[7] Arabian-Hoseynabadi H, Oraee H and Tavner P J 2016 *International journal of electrical power and energy systems* **32** 817–824
[8] Rafiee J, Arvani F, Harifi A and Sadeghi M 2007 *Mechanical Systems and Signal Processing* **21** 1746–1754
[9] Li Y, Le C, Ding H, Zhang P and Zhang J 2019 *Journal of Marine Science and Engineering* **7** 115
[10] Jaiswal V and Ruskin A 2019 *OTC Offshore Technology Conference* **1** 1–7
[11] Martinez-Luengo M, Kolios A and Wang L 2016 *Renewable & Sustainable Energy Reviews* **64** 91–105
[12] Joshuva A D, Aslesh A and Sugumaran V 2019 *International Journal of Mechanical Sciences* **9** 95–112
[13] Gorostidi N and Nava V 2021 *Book of Proceedings of the 17th EAWE PhD Seminar* **1** 149–156
[14] Antoniadou I, Manson G, Staszewski W, Barszcz T and Worden K 2015 *Mechanical Systems and Signal Processing* **64-65** 188–216
[15] Laboratory N R E 2021 Openfast documentation [Online; accessed 10-February-2022] URL `https://openfast.readthedocs.io/en/main/`
[16] Robertson A, Jonkman J, Masciola M and Song H 2014 Definition of the semisubmersible floating system for Phase II of OC4 Tech. Rep. 1 National Renewable Energy Laboratory (NREL) Denver, CO
[17] Lee S, Churchfield M, Driscoll F, Sirnivas S, Jonkman J, Moriarty P, Skaare B, Nielsen F G and Byklum E 2018 *Energies* **11**

Time domain hydrodynamics at panel level: an application to the structural analysis of Telwind

E Rodríguez-López, C Moreno-Narrillos, F Rueda-Guglieri, Á Yáñez-González
ESTEYCO, Avda. Burgos 12B-Bajo, 28036, Madrid, Spain

E-mail: eduardo.rodriguez@esteyco.com

Abstract. The current modelling approaches for Floating Offshore Wind Turbines (FOWT) present some limitations in the representation of the fluid-solid interaction. Mainly, they do not consider full flexibility of the floating body, so internal forces on the platform cannot be resolved. ESTEYCO has developed a novel methodology to couple all the actions (hydrodynamics, structural, aerodynamics, control, etc.) that are applied over a FOWT concept. The main novelty of this approach resides in selecting Abaqus (well stablished Finite Element –FE– tool) as the central solver of the motion equations, naturally enabling inclusion of full platform flexibility (and not only rigid bodies linked by flexible members). The inclusion of the coupled hydrodynamic actions (highly dependent on platform movement) is performed through a novel tool: EST-PANEL which provides the time domain hydrodynamic pressure at panel level. This approach is thoroughly validated against other well established modelling techniques showing outstanding agreement both in platform movement and internal tension of cables. This methodology has been developed to support the design and development of ESTEYCO's own technology: Telwind. The current approach thus provides a unique fully coupled FE tool to perform a detailed structural assessment of the floating platform even from the earliest conceptual stages of the design. In addition, the current approach can be easily extended to simulation of some loading scenarios (cable breaking, transient deployment, non-linearities) that are not easy to model with other simulation tools.

1. Introduction

Due to the increasing demand for offshore wind power production (more than 230 GW are required to achieve carbon neutrality in Europe by 2050: Ref. [1]), Floating Offshore Wind Turbine (FOWT) concepts are unavoidable, especially in countries, such as Spain, where deep coastal waters require floating solutions. To support the design and analysis of new FOWT concepts, coupled analyses are performed as recommended by design standards such as Ref. [2], where the dynamic loads and responses of the system are calculated by simultaneously considering all the relevant excitations (wind, waves, current, actuation, etc.).

Several examples of simulation codes are reviewed in Ref. [3]. These tools are usually developed with a high computational efficiency (to perform several simulations under multiple conditions) by considering some modelling simplifications. Although these simplifications may be sufficient for most design justifications, certain aspects of the design may require a more detailed structural analysis of the floating structure.

From the perspective of designing the floater concept, one of the main limitations is the representation of the fluid-solid interaction and the calculation of hydrodynamic loads. Different simulation codes introduce various levels of abstraction which, at most, consider several rigid bodies joined by flexible members [4], or a fully flexible structure as far as it can be represented by 1D beam elements. However, this approach may not be suitable for all floating platform concepts: firstly, the platform design may not be suitable for a 1D beam element representation; secondly, the instantaneous pressure loads on the wet surface may not always be calculated during coupled simulations; thirdly, it does not enable resolving the internal forces on the platform other than at joining members. The latter point is significantly relevant since in fact, for the calculation of the internal forces transmitted by flexible members, the flexibility of otherwise rigid bodies needs to be considered if the system is hyperstatic.

In the current work, a novel methodology is being developed to integrate, in a fully coupled manner, all the actions (hydrodynamics, structural, aerodynamics, control, etc.) that are applied over a FOWT concept with the main focus on the structural behavior of the floating platform. This novel approach implements a time domain solver of the structural and linear hydrodynamic equations of the system in the FE software Abaqus. The hydrodynamic loads (added mass, radiation, diffraction and restoring forces) are applied as distributed pressures over the wet surface. The main novelty of the current approach resides in selecting the FE software Abaqus as the main integrator of the structural and dynamics of the platform. This implementation solves the three main limitations indicated earlier thus allowing a detailed structural assessment of the floating platform and any of its components. In fact, the current approach presents a higher versatility and enables implementation of certain situations and modelling techniques (non-linearities, transient scenarios, substructuring, lower tank deployment, cable breaking, etc.) which are not easy to consider with other modelling approaches. Further work is underway to link Abaqus to certain modules of OpenFAST in order to fully couple the aeroelastic and control actions.

The paper includes in Section 2 a description of the newly developed methodology and the test cases simulated for its validation against OpenFAST using as a test case Esteyco's Telwind platform concept. Section 3 presents the results of the validation performed considering both a rigid body platform and a fully flexible platform. Section 4 includes the conclusions of the work.

2. Methodology

2.1. Modelling approach in Abaqus

The Telwind floating platform is formed by two bodies (upper and lower tank) joined by a series of tendons (see geometry in Figure 2). The floating platform is modelled by means of a flexible Finite Element (FE) discretization in Abaqus [5] which resolves the deformation of the floating platform along with its motions. The structural behavior is simplified as a deformable shell (although any other type of representation may be implemented) where the Young's Modulus is adjusted to mimic the rigidity provided by internal stiffeners not included into the model. The tendons joining the two floating bodies are modelled as axial connectors without mass or inertia. The mass, and inertia of the floating bodies are adjusted where required by adding non-structural mass to ensure an exact match to the OpenFAST model, thus ensuring a consistent validation. The tower is modelled as a 1D beam whose mass and inertia properties are matched to the correspondent OpenFAST simulation but with infinite rigidity.

The external skin of the FE model coincides with the hydrodynamic surface. Hydrodynamic actions are strongly dependent on the platform motion so cannot be precomputed. To solve this, ESTEYCO has developed EST-PANEL, a novel tool that enables resolving the hydrodynamic actions over the floating body. The method is summarized in Figure 1 and described in detail in Section 2.1.1.

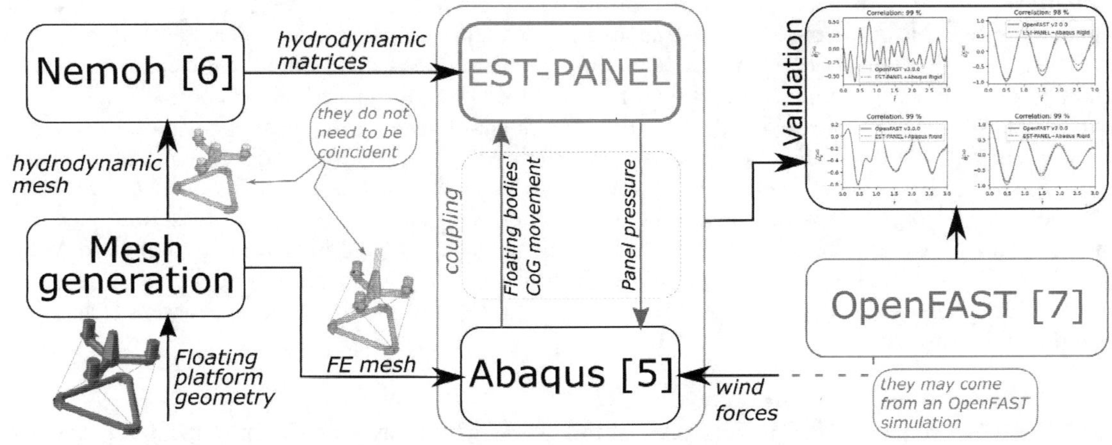

Figure 1. Summary of methodological approach.

2.1.1. Time domain application of hydrodynamics at panel level

The linear hydrodynamic matrices are computed in Nemoh [6] using a panel discretization (which does not need to be the same as the structural FE discretization) and linear first-order potential theory. Second-order wave loads and viscosity effects are not considered. For cases where, unlike the current study, the mesh discretization is not the same, linear interpolation based on centroid position may be enabled. These matrices provide the pressure at each panel in the frequency domain as a function of the bodies' movements (for radiation) and wave spectrum (for excitation forces). These pressures are converted into the time domain (added mass, A_∞, Cummins' kernel, K, and wave excitation, P^{exc}_{wave}) by means of an in-house Python tool. The hydrostatic pressure, P_{hst}, is imposed based on the instantaneous position of the FE element. The wave excitation pressures are multiplied by a given realization of wave spectrum with arbitrary phases taken from an equivalent Hydrodyn [7] simulation, so the comparison with OpenFAST includes the same wave history. The actual wave pressure over each panel, P^i_{wave}, is therefore calculated as a function of time, but independent of bodies' movement.

Note that the global buoyancy force is obtained by the sum of each panel's hydrostatic pressure, so, at the initial instant, it equilibrates the action of gravity. As the body moves (and deforms) the hydrostatic pressure is recomputed at each time step and the resultant force changes accordingly with time/position of the body removing the need to prescribe a linear hydrostatic matrix.

EST-PANEL, (written in Fortran) is coupled to Abaqus through a series of user subroutines and, for each time step, computes the applied pressure, P^i_{panel}, of each FE element, i, as a function of the instantaneous bodies' displacement, x, velocity, $\dot{x}$, and acceleration, $\ddot{x}$ (τ represents a dummy time lag to compute the integral):

$$P^i_{panel}(t) = A^i_\infty \ddot{x}(t) + \int_0^t K^i(t-\tau)\dot{x}(\tau)\mathrm{d}\tau + P^i_{wave}(t) + P^i_{hst}(x(t))\,.$$

One of the main advantages of the current approach is the ability to simultaneously resolve the platform global motions and the stress/deformation field. The integral of all the external actions gives the resultant body force which generates the rigid-like motion. Further, the structural deformation caused by these actions is also computed and, consequently, the stress field is simultaneously obtained.

2.1.2. Simplifications of the current implementation

The main focus of the current study is the validation of the proposed approach where the hydrodynamic actions are modelled as pressures applied over the bodies' surface fully coupled with the platform deformation and movement computed in Abaqus. Consequently, some modelling simplifications are introduced in the model to facilitate the validation process and avoid introducing unnecessary complexity in the FE model that can add uncertainty to the validation results.

The mooring system action is replicated as simple linear springs attached to the upper floating body and additional linear damping (proportional to the velocity) is introduced at the bodies' CoG. These two effects are exactly replicated in the OpenFAST simulation. Sea current is not considered.

Wind loads are precomputed by an equivalent OpenFAST simulation and subsequently applied as a time-varying follower point load at the tower top node. Note that this is only valid if the resultant displacement history from both simulations is equivalent (which will be shown in Section 3.1). The lack of coupling with the aerodynamic actions is detected as a potential improvement point and works are currently conducted to include a modification of OpenFAST to enable fully coupled simulations that solve the aerodynamic and control problems as a function of the displacements and deformations computed by Abaqus.

2.2. *Modelling approach in OpenFAST*

OpenFAST [7] version 3.0.0 has been used in this work, which has been recently updated to include multiple floating bodies and account for the flexibility of the floating platform members. The SubDyn module is thus used to model the two floating bodies (rigid bodies) by means of six cables. No modal reduction (Crag-Bampton) is performed in such a model. The two floating bodies (Upper and Lower tanks) are each modelled using potential flow theory in the HydroDyn module, which accounts for hydrodynamic loads. The hydrodynamic loads on the cables are neglected. For the validation test cases, the floating platform is considered in OpenFAST as a single rigid body, and thus modelled using a single body defined by potential flow theory.

Linear stiffness and damping matrices are added in the platform reference point to model the mooring and additional linear damping (consistent with the simplified representation in Abaqus). The tower is modelled as fully rigid. Only the mass and inertia are represented to match those of the Abaqus model.

The WTG and the control considered for DLC test cases [2] are taken from the DTU 10 MW reference wind turbine [8].

2.3. *Test cases*

Test cases are selected to validate the new analysis methodology developed in Abaqus against the well accepted software OpenFAST. These cases are shown in Table 1 and include: i) free decay simulations, to validate the correct implementation of the Cummins equation and hydrostatic loads (no wind or waves are present); ii) DLC 1.6 and DLC 6.1 with no wind, to validate the correct implementation of the wave loads (assuming a JONSWAP spectrum); and iii) DLC 1.6 and DLC 6.1 with wind, to evaluate the possibility to include wind loads as a time history obtained directly from the OpenFAST simulation and assess the need for a fully coupled simulation.

For the validation cases the tendons and the bodies are taken as a rigid solid so its movement can be directly compared with a single rigid body (with 6 DoFs) simulated in OpenFAST

Turbulent wind is generated using the external tool TurbSim [9] with the IEC 61400-3 standard [2], and used for the OpenFAST simulations. Both wind profiles are computed with an expected value of turbulence intensity at 15 m/s Iref=0.16 and a turbulence scale Λ=42 m. Normal Turbulence Model is employed for the DLC 1.6, giving σ_u=1.856 m/s (23.2% of mean wind speed at hub height); Extreme 50-Year Wind Speed Model is used for DLC 6.1, providing σ_u=5.5 m/s (11% of mean). As mentioned, the wind loads are then applied in the Abaqus model as a concentrated (point force at the tower top) force. This aspect of the methodology is being improved in a parallel project.

Table 1. Test cases used for the validation of the Abaqus methodology.

Case	Initial conditions	Wind	Waves
Heave free decay	Unitary heave displacement	-	-
Pitch free decay	Unitary pitch rotation	-	-
DLC 1.6	Initial transient removed	No wind 8 m/s	Irregular waves[a]
DLC 6.1	Initial transient removed	No wind 50 m/s	Irregular waves[b]

[a] Hs = 5.3 m, Tp = 7.7 s
[b] Hs = 10.1 m, Tp = 17 s

2.4. Model results

The implementation of EST-PANEL enables saving the different contributions of the hydrodynamic pressure at different time instants. A visualization of the pressure fields for an arbitrary time instant can be observed in Figure 2. The wave pressure is about one order of magnitude higher than the other two examples. The time instant corresponds to the peak of a wave in a DLC 6.1 (no wind) case. The added mass component reflects the downwards movement (positive and negative pressure in the lower and upper surfaces respectively).

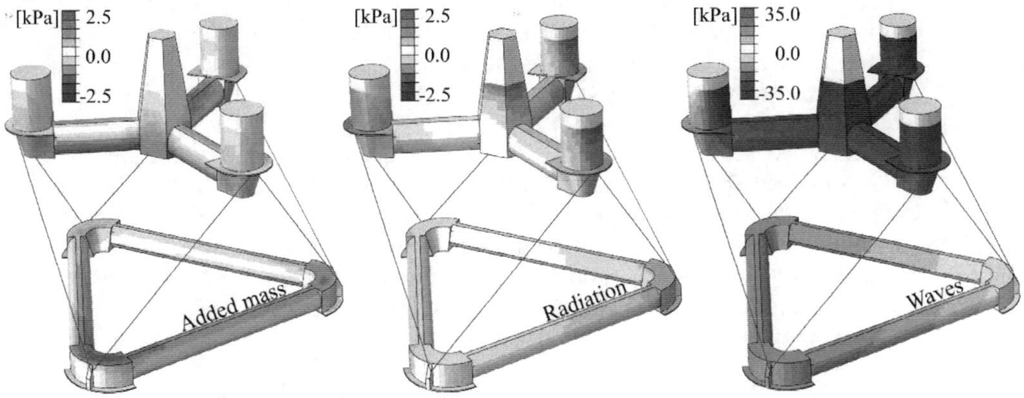

Figure 2. Different components of hydrodynamic pressure field.

In order to check the correct implementation of the FE model and have some preliminary insight on its behavior, the main structural modes can be studied. Figure 3 shows the main two modes related to cable flexibility. The first one represents a swing motion of the lower body coupled with a lateral movement of the upper body. The second mode represents a relative displacement in the vertical direction of both bodies. The second mode appears at roughly twice the frequency of the former. In these two cases, the structure of the body remains almost undeformed since the cables are more flexible than the structure. This will have some impact on the results where the main contribution to the relative movement and flexible deformation will be originated by the cables' deformation.

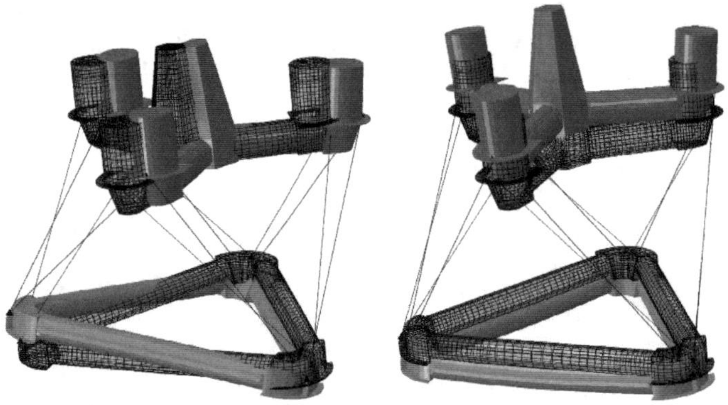

Figure 3. Main two structural modes related with cable flexibility.

3. Results and discussion

The validation of the methodology is performed in two stages. First, the methodology is validated considering the floating platform as a rigid body both in Abaqus and OpenFAST. Heave and pitch motions of the global centre of gravity of the platform are compared for this validation.

Second, the methodology is assessed assuming a fully flexible floating platform (Abaqus) and including the cable flexibility between the two floating bodies (OpenFAST). Heave and pitch motions of the centre of gravity of each floating body are compared in this simulation, as well as the cable tension.

3.1. Methodology validation against OpenFAST simulations

Normalized heave and pitch motions of the platform centre of gravity are compared between Abaqus and OpenFAST for free decay (Figure 4), a representative DLC 1.6 (Figure 5) and a representative DLC 6.1 (Figure 6).

The comparison of the heave and pitch free decays in Figure 4 show an excellent correlation between the results obtained with OpenFAST and the new methodology implemented in Abaqus. The decaying periods are matched almost perfectly, as well as the amplitude of the pitch oscillation. The amplitude of the heave oscillation seems to be slightly more damped in the simulation in Abaqus, which may be related to the numerical damping of the FE integration scheme. These results validate the correct implementation of hydrostatic, added mass and radiation memory loads.

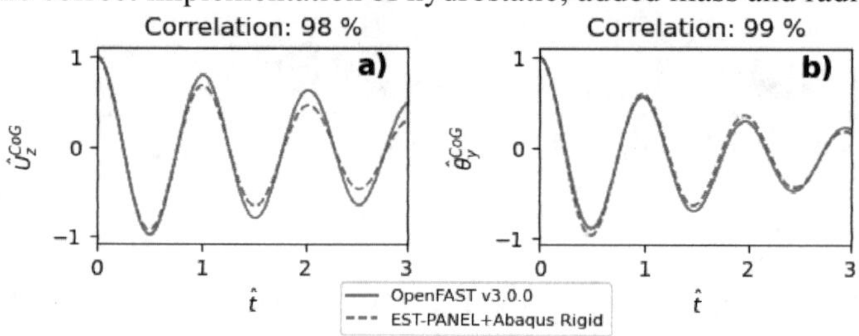

Figure 4. Comparison between normalized platform motions calculated with OpenFAST and EST-PANEL + Abaqus for (a) heave and (b) pitch free decay simulations. Correlation between OpenFAST and Abaqus simulated responses is included.

The simulation results for the representative DLC 1.6 in Figure 5 also show a remarkable correlation between the normalized platform motions calculated using OpenFAST and Abaqus. In the comparison of the movements obtained when no wind is included in the simulation (i.e., only wave loads are considered), the two curves (OpenFAST and Abaqus) cannot be discerned. This result validates the correct implementation of the wave loads in the new Abaqus methodology.

When wind is included in the simulation, the correlation is still excellent, although a small shift in the heave platform motion can be seen in Figure 5. The marginally worse agreement suggests that the current consideration of wind loads at the tower top as a force time history obtained from the OpenFAST simulation output may introduce some error in the Abaqus simulation since it is not fully coupled with the platform and rotor motions (see Section 2.1.2). This effect is however negligibly small for the wind speed of 8 m/s considered in this test case.

The addition of wind load in this test case does not greatly change the relative platform motions with respect to the case without wind. This is especially true for the heave motions, which is practically the same with and without wind. The pitch variations are similar between the two simulations but a constant offset of approximately 3° is observed for the case with wind due to the mean aerodynamic action over the rotor tilting the whole system.

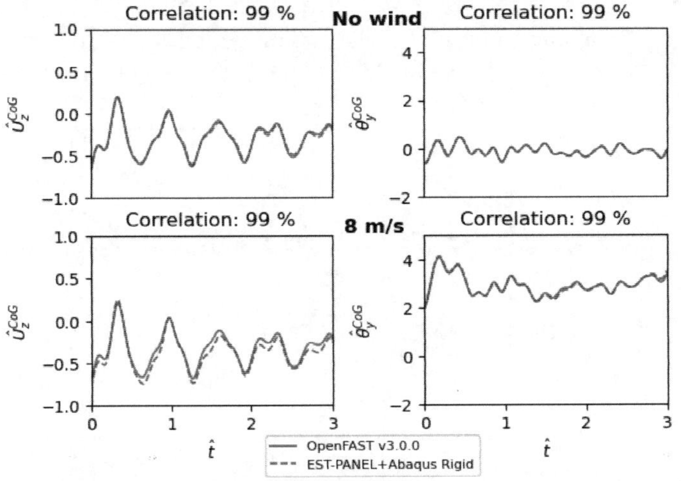

Figure 5. Comparison between normalized platform motions (heave and pitch) calculated with OpenFAST and EST-PANEL + Abaqus for a representative DLC 1.6 considering: (top) no wind loads and (bottom) representative wind at 8 m/s. Correlation between OpenFAST and Abaqus simulated responses is included.

The simulation results of the DLC 6.1 test case, depicted in Figure 6, show good correlation for pitch motions (97%), whereas some differences (both in amplitude and frequency response) appear for the heave motion in the simulations without wind (92% correlation) and with wind loads (79%).

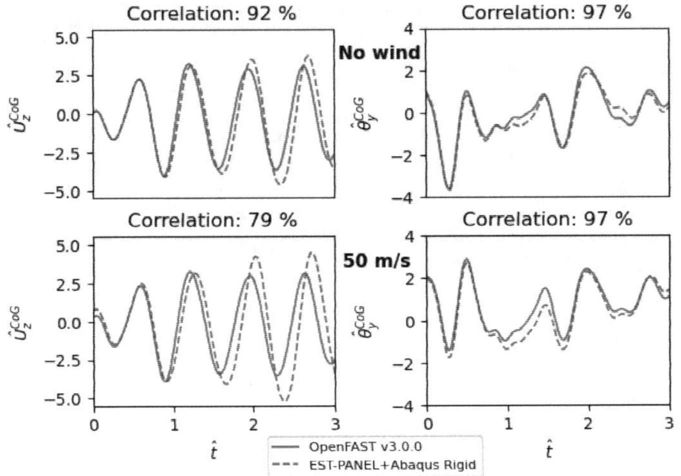

Figure 6. Comparison between normalized platform motions (heave and pitch) calculated with OpenFAST and EST-PANEL + Abaqus for a representative DLC 6.1 considering: (top) no wind loads and (bottom) representative wind at 50 m/s. Correlation between OpenFAST and Abaqus simulated responses is included.

This test case considers an extreme sea state which results in larger platform motions. Differences in the simulation results may arise from the small displacement assumptions made by each software. For the EST-PANEL + Abaqus case, the hydrodynamic action is applied at each time instant over the displaced body position (geometric nonlinearity is accounted for in Abaqus), so the force resultant may differ from that computed by OpenFAST, which assumes small displacements around the equilibrium position. It is reasonable to expect this discrepancy to be greater for larger platform displacements in extreme sea conditions. When including wind loads, the incomplete treatment of such wind loads is more evident, likely due to the differences in displacement of the application point. These differences in displacement should be solved by coupling the aero-elastic part of OpenFAST within the Abaqus dynamic simulation.

3.2. Flexible body calculations

3.2.1. Comparison of results against OpenFAST

The same test cases simulated in Section 3.1 considering the floating platform as a rigid body are simulated in this section including cable flexibility in OpenFAST and a fully flexible platform in Abaqus. The results in Figure 7 show the heave and pitch motions of the Upper and Lower tank (UT and LT) centre of gravity for the three simulation cases considered: free decays, DLC 1.6 and DLC 6.1, respectively.

The results are nearly equivalent using both simulation tools. The natural periods of the platform motion (free decays in Figure 7) slightly increase with respect to the rigid body case (Figure 4), as can be expected from a more flexible body.

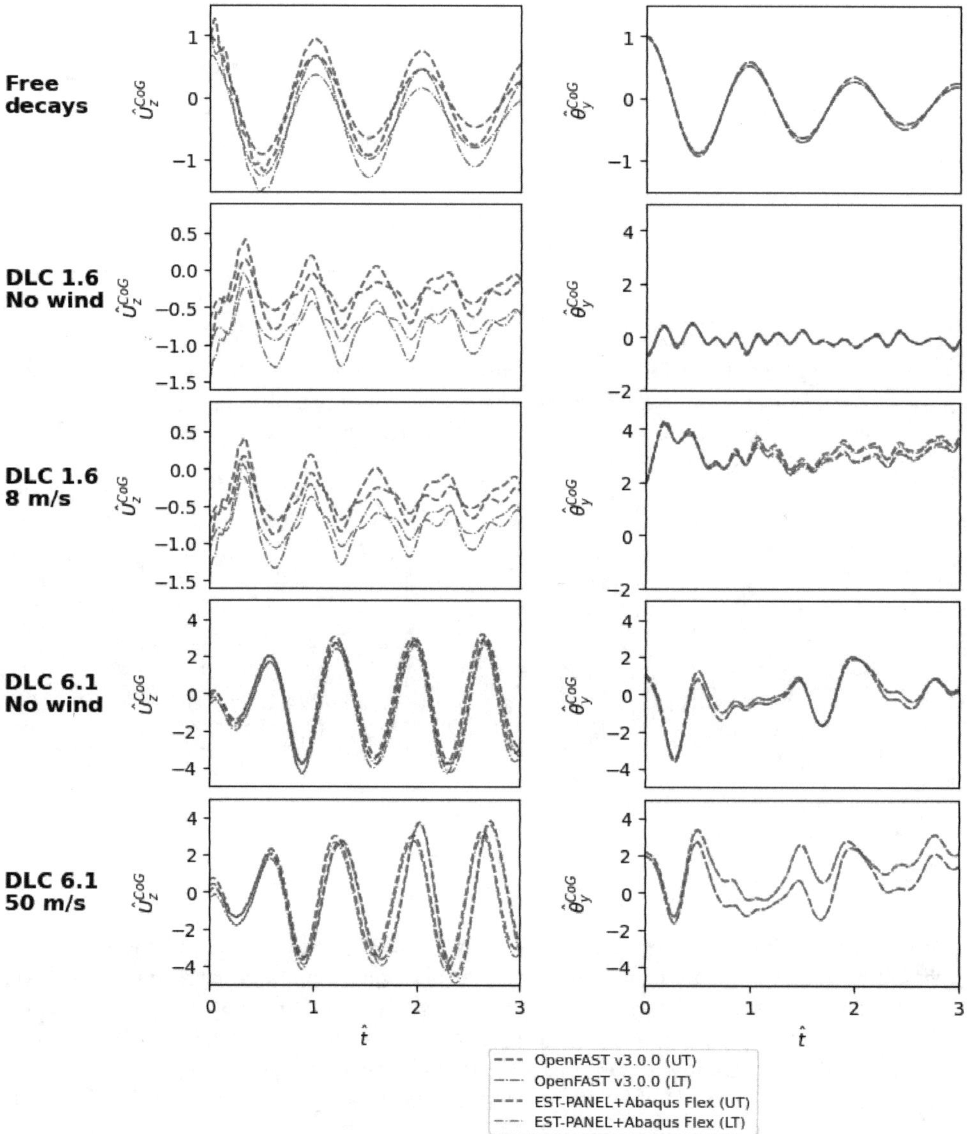

Figure 7. Comparison between normalized platform motions (heave and pitch) calculated with OpenFAST and EST-PANEL + Abaqus for different loading scenarios (Table 1) considering two hydrodynamic bodies joined by flexible cables.

Initial tension on the cables is assumed to be zero so equilibrium is achieved during the initial transient of the simulation. This generates a vertical separation between the two bodies due to the elongation of the cables caused by the gravity loads which can be clearly seen in Figure 7. The average value of vertical displacement of the LT is the same in OpenFAST and Abaqus since the cable stiffness defined in both simulations is the same. However, the amplitude of the oscillations is slightly larger in OpenFAST. This is mainly caused by the treatment of the initial transient: on the one hand, the simulation with Abaqus includes an initial static step that allows to achieve equilibrium (mass and floatability) before the dynamic simulation is performed. On the other hand, OpenFAST does not allow this initial static step, so equilibrium needs to be reached during the simulation, implying that the initial displacement for the free decays introduces a larger potential energy since the initial position of the Lower Tank does not include initial cable stretching. Additionally, the different consideration of damping (see comments on Figure 4) may also affect these amplitude differences.

The differences in the amplitude oscillation are relatively small compared to the large platform motions in the DLC 6.1 shown in Figure 7. In this DLC 6.1, the differences in the heave and pitch motions between OpenFAST and Abaqus are negligible, especially in the case in which wind loads are neglected. It is interesting to compare the results for the DLC 6.1 with those obtained considering a rigid body platform in Figure 6. The agreement between OpenFAST and Abaqus appears to be better when a flexible body is considered in the simulation.

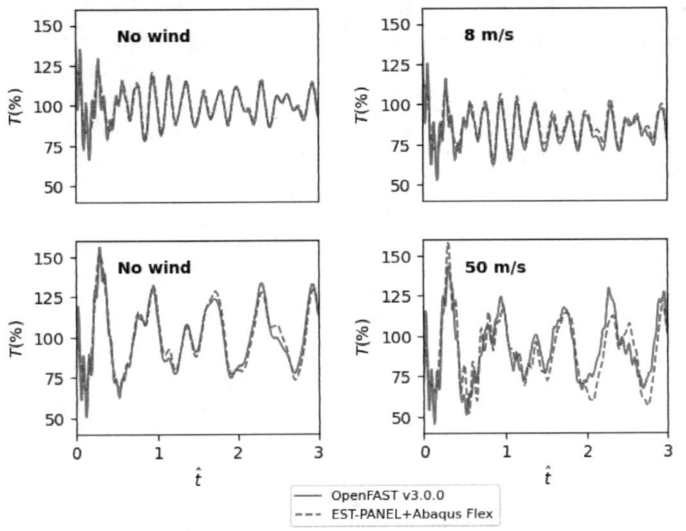

Figure 8. Comparison between the normalized tension of one cable calculated with OpenFAST and EST-PANEL + Abaqus considering two hydrodynamic bodies joined by flexible cables for two representative DLCs: (top) DLC 1.6 (bottom) DLC 6.1.

The tension (normalized to the static cable tension in equilibrium) on one of the tendons that link the two floating bodies is shown in Figure 8 for the two DLC cases simulated both with and without wind loads. The agreement in the results obtained with the OpenFAST and Abaqus simulations is excellent, and only some minor differences appear for the DLC 6.1 that considers wind loads. The slightly larger oscillations in cable tension during the initial stages in OpenFAST are mainly due to the differences on the initial transient since it does not include an initial static step where equilibrium is reached. The transient effect on the oscillations reduces with time, so consistent comparisons can be performed after this effect disappears. This result further validates the novel EST-PANEL + Abaqus approach against OpenFAST. However, the Abaqus simulation would allow a greater versatility in the modelling of the floating platform, e.g., include non-linearities such as cable slacking or material constitutive law, deployment of the lower tank, breakage of a tendon, impact loads, etc.

4. Conclusions and future perspectives

The current work presents and validates a new methodology to perform coupled simulations of FOWT that enables modelling floating platforms as fully flexible bodies with the main focus on their structural behavior. The novel approach implements a time domain solver of the equations of the system's motion in the FE software Abaqus. The hydrodynamic loads (added mass, radiation, diffraction and restoring forces) are applied as distributed pressures over the surface by means of EST-PANEL, a novel tool (fully coupled at each time step with Abaqus) that calculates the different components of the hydrodynamic pressure (assuming linear potential theory) at each element of the structure's discretization.

The methodology has been thoroughly validated against OpenFAST v3.0.0 which is considered amongst the reference software for FOWT simulation. Different representative test cases are compared (free-decay tests, DLC 1.6 and 6.1) with and without incoming wind. The platform vertical motions are compared both enabling or not the full platform flexibility. In both cases, there is a remarkable agreement between the current novel approach (EST-PANEL + Abaqus) and the reference software. Furthermore, for flexible cases, the cable tension may also be compared. The agreement is also significantly satisfactory even provided OpenFAST does not enable to compute an initial static step to reach the initial equilibrium. For both rigid and flexible cases, slightly larger differences (but well within admissible discrepancies) appear for large excitation situations (DLC 6.1) where the different treatment of small displacements may play an important role.

Whereas the current approach can be considered as fully validated since its satisfactory performance has been shown for a broad variety of representative test scenarios, the full potential of this tool is well beyond what has been shown in the current study. The main novelty of the current approach resides in selecting the FE software Abaqus as the main integrator of the structural and dynamics of the platform. One of the main reasons to do that was to enable a straightforward implementation of platform flexibility. However, by doing this, some of the current limitations of conventional FOWT simulation software can be easily overcome such as: consideration of non-linearities (material constitutive law, load following application point, cable slacking, contacts between different bodies...), implementation of any type of structural element (shell, solid), application to transportation and installation phases, accidental scenarios –cable breaking–, or lower tank deployment, amongst many others.

This methodology can be significantly useful especially during the preliminary phases of the design by performing a limited number of pre-selected Design Load Cases (DLCs) driving the design that need to be more deeply studied than what is currently possible with conventional FOWT simulation software. Although computational time of the EST-PANEL + Abaqus is higher than OpenFAST (by a factor that may range from 5 to 20 depending on several variables), it is still affordable to simulate some design-driving DLCs with higher loads.

In addition, further progress in this approach is currently being developed. One of the weakest points was the inclusion of wind loads as a precomputed external force. A modification of some OpenFAST modules is ongoing in such way that the aerodynamic and aeroelastic actions as well as wind-turbine control can also be coupled with the current integration of the platform motion (and flexibility) in Abaqus. Another ongoing development is the inclusion of a realistic mooring system based on truss elements.

Acknowledgements
This work has been funded by CDTI (Centro para el Desarrollo Tecnológico Industrial, Spain) through the project LEAF (Liderazgo Español para al Avance de la eólica Flotante) and supported by the Ministry of Science, Innovation and Universities, Spain.

References
[1] WindEurope 2019 *Our energy, our future. How offshore wind will help Europe go carbon-neutral*
[2] IEC 61400-3 *Wind Turbines – Part 3: Design requirements for offshore wind turbines*
[3] Robertson A et al. 2014 Offshore code comparison collaboration continuation within IEA wind task 30: Phase II results regarding a floating semisubmersible wind system *Proc. Int. Conf.*

Offshore Mech. Arct. Eng. - OMAE vol 45547 (American Society of Mechanical Enineers) p V09BT09A012
[4] Jonkman JM et al. 2019 Substructure Flexibility and Member-Level Load Capabilities for Floating Offshore Wind Turbines in OpenFAST *Proc. Int. Conf. Offshore Mech. Arct. Eng. – OMAE* vol 59353 (American Society of Mechanical Enineers) p V001T01A004
[5] Simulia 2017 *ABAQUS User's Manual v2017*
[6] Babarit A and Delhommeau G 2015 Theoretical and numerical aspects of the open source BEM solver NEMOH *Proc. of the 11th European Wave and Tidal Energy Conference (EWTEC2015)*
[7] Jonkman J and Sprague M *OpenFAST v3.0.0* https://www.nrel.gov/wind/nwtc/openfast.html (accessed October, 2021)
[8] Bak C et al. 2013 The DTU 10-MW reference wind turbine *Danish wind power research*
[9] Jonkman B J and Buhl Jr. M L 2006 *TurbSim user's guide. No. NREL/TP-500-39797* National Renewable Energy Lab.(NREL), Golden, CO (United States)

Journal of Physics: Conference Series 2257 (2022) 012010 doi:10.1088/1742-6596/2257/1/012010

Validating the next generation of turbine interaction models

T Levick[1], A Neubert[2], D Friggo[3], P Downes[1], R Ruisi[1] and J Bleeg[1]

1. DNV, One Linear Park, Avon St, Temple Quay, Bristol BS2 0PS, UK
2. DNV, Emsstraße 18 c/o Glaswerk, 26135 Oldenburg, Germany
3. DNV, Av. Carlos Gomes 300, sl. 1102 (Boa Vista), Porto Alegre, Brazil

Tom.Levick@dnv.com

Abstract. It is important to validate turbine interaction models to understand the uncertainties and biases inherent when we model wind farm power output for future wind farms. We present here a repeatable and model-agnostic methodology developed for validating wind farm production models. Power data from the Supervisory Control and Data Acquisition systems of wake-free turbines are used with turbine power curves to generate inlet wind speeds representative of average conditions on the front row of a wind farm. These wind speeds are used, with other model inputs, to run models and predict a modelled power time series for each turbine. The modelled and measured power time series are compared to derive mean bias error metrics. The methodology is applied at 6 offshore wind farms to test established and novel turbine interaction models. We compare the distributions errors predicting power at turbines across models and wind farms. We find that the new models, CFD.ML and the Stratified Eddy Viscosity model, perform well with respect to the established WindFarmer Eddy Viscosity model, and see increased errors for the largest wind farms. We discuss methodological uncertainties in the input wind speed derivation that may cause biases in the overall distributions at windspeeds near the turbine low wind speed cut-in and rated power, and make suggestions for future methodological refinements.

1. Introduction

Quality-controlled wind farm Supervisory Control and Data Acquisition (SCADA) data provides a source of measured turbine power time series that can be used to validate turbine interaction models. Through work analysing 7 GW of offshore operational wind farms, DNV has access to 4000 turbine-years of quality-controlled SCADA data. This dataset presents an opportunity to derive valuable insights into the efficacy of turbine interaction models used in pre-construction wind energy assessments. However, there are significant challenges in producing a reliable turbine interaction model validation using production data. In this study we share a repeatable validation methodology which is scalable across large collections of operational datasets. We expose and mitigate some methodological uncertainties in the process. We also present results for a subset of offshore operational sites representative of different farm scales and for a range of established and novel flow and wake modelling approaches.

The need for new turbine interaction model validations.

Some wind farms have been widely used to validate models, such as the Horns Rev I and Nysted offshore wind farms used in [1], [2] and [3], leading to a potential risk of model over-fitting. As further offshore wind farms are built, we start to see large wind farm clusters and many of today's new offshore wind farms have a significantly greater number of wind turbines than those considered within historical turbine interaction validation studies. Today's offshore wind turbines are getting larger, with greater rotor-diameter-to-hub-height ratios, compared with turbines at historical validation sites. We should consider modern wind farms in updated validations.

A model's prediction of the turbine interaction loss and its associated uncertainty have a significant bearing on the wind farm project value. As understanding of the turbine interaction effect has increased, Energy Production Assessments (EPAs), conducted before farm construction, have incorporated various correction factors, for example the large wind farm / deep array correction [3], or a correction to account for blockage [4]. By building a better understanding of turbine interactions we can reduce unexpected future project underperformance relative to EPAs and, in providing a validated demonstration of reduced modelling uncertainty, we can enable a reduction in EPA uncertainty and thus also the cost of capital for wind farm investments.

If validation helps us to better understand turbine interactions, we can use such knowledge to build more reliable models. The improved models can be used to inform more optimal wind farm design decisions. With the introduction of wind farm control strategies such as wake steering through active yaw control, this understanding will be crucial for designing control strategies. Levelised cost of energy could be reduced by using our improved turbine interaction models to develop wind farm configurations with increased energy output, and more effective utilisation of wind farm assets.

1.1. A time domain framework for turbine interaction model validation?

Atmospheric stability is time dependent with diurnal and seasonal cycles exhibited at many wind farm sites. Turbine interactions are stronger in stable atmospheric conditions [5], [6]. The wind speed and direction frequency distribution can vary by season and time of day. Given that turbine interaction losses are extremely sensitive to wind direction, both these factors can introduce time dependency in turbine interaction losses. As a result, it is possible that models also exhibit time-dependent modelling errors conditioned on their abilities to model the response to these drivers.

Today's wind projects must compete and sell energy at market price with little or no subsidy support. The market price for energy is highly time-dependent with seasonal and diurnal patterns. When combined with turbine interaction losses exhibiting similar patterns it is possible there may be correlating factors. This could cause impacts on the central estimate of project revenue that do not exactly cohere with the central estimate of annual energy production.

Furthermore, turbine power performance is an inherently non-linear process and is becoming more difficult to represent within traditional frequency domain methods. This is because pro-active wind farm operational control drives increasingly complex dynamic turbine rating and curtailment policies.

As new models emerge to solve new industry demands, it is advantageous to have a generalised validation framework within which to test multiple types of models. Models with greater dependencies can compete with simpler models over the same time periods. For these reasons, we generate power time series from all models for direct comparison with measured power time series data, and reserve binning and aggregation of modelling errors to later steps.

1.2. The challenge of predicting free-stream wind speeds from SCADA data

Within EPAs, on-site measurements are typically used to define the prevalence of wind conditions. These wind conditions are used as input to turbine interaction models and to drive predictions of power and annual energy production. The models require the input of free stream wind speeds, that is, the wind speeds in the absence of any wind turbines. These wind speeds are typically either measured offshore by a Floating LiDAR Device (FLD) or a meteorological mast, or predicted using an atmospheric model such as the Weather Research and Forecasting model (WRF).

To validate any given turbine interaction model, measurements of free-stream wind speeds concurrent with the SCADA data from an operational wind farm would be preferable. Unfortunately, reliable free stream wind speed measurements are generally not available post wind farm construction. Instead, it is common to validate models using only the turbine SCADA data. The turbine power signal can be converted into a wind speed using manufacturer provided turbine performance information, i.e., the power curve. This is performed for one or more selected turbines in the wind farm. The turbines selected are usually those most upstream in the wind farm for each wind direction (the "front row" turbines), ensuring we exclude wake impacted turbines. With front-row wind speeds as input, a turbine interaction model is used to predict wind speeds incident at the other turbines, then power is derived using the same power curve information. To compare models and measurements, studies typically normalise production to the front row of the wind farm.

The turbine interaction loss is defined as the difference in production between a turbine in the presence of a wind farm and what it would produce in isolation, that is, as if all other wind turbines were absent. A focus on wake validation studies where power predictions are normalised to front row production has meant the impact of blockage has been largely overlooked in popular models used for Energy Production Assessments. In [4] it is shown that the front row of the wind farm experiences lower wind speeds than an individual turbine on that front row would expect to experience in isolation. The methodology presented in this validation study cannot be used to measure the blockage component of the turbine interaction loss. However, by considering error metrics at a turbine level in our validation we expect to see benefits from models which capture effects such as a power variation across the front-row. We also see significant value in testing how models perform relative to front-row conditions.

2. Validation methodology

The validation methodology can be summarised as follows, building on work first presented in [7]. Firstly, SCADA data from offshore wind farms are compiled to form a validation database representing measured turbine production. For each wind farm, front-row-average wind speed time series are derived from selected "front row" turbines using the SCADA power data, and wind direction also comes from the turbine yaw angle after a quality control process. Due to the lack of suitable on-site measurements, data from the New European Wind Atlas (NEWA) [8] or Vortex SERIES [9] are used to define atmospheric stability. These data are used to define time stamped 10-minute flow cases which are used as input to each prediction of a modelled power time series.

Further filters are applied to both measured and modelled time series to ensure methodological uncertainties are reduced. These filtered measured and modelled time series are compared, and modelling error metrics are derived, providing consistent high level model comparisons across the considered wind farms.

2.1. Wind farms considered for validation

The validation database currently consists of 6 offshore wind farms representing a variety of wind farm scales as summarised in **Table 1**. The wind farms are drawn from DNV's commercial wind farm operational data sets and, for confidentiality, are referred to as Small, Medium, Large and Extra-Large (S, M, L, XL), according to the number of turbines within the wind farm. At each wind farm at least one year of 10-minute SCADA power data was available.

Table 1. Summary of wind farm data considered in this study

Farm	Abbreviated name	Number of turbines	Data period in years	Raw data 10-minute records	Valid 10-minute records	NEWA data available?	Flow Model available?
Small	S1	< 25	4.5	138559	75578	Yes	No
Medium	M1	25 to 75	1.3	32871	18637	Yes	No
Large 1	L1	75 to 125	2.5	47573	24682	Yes	No
Large 2	L2	75 to 125	3.2	139382	35279	No[1]	Yes
Extra Large 1	XL1	> 125	1.0	21445	5631	Yes	Yes
Extra Large 2	XL2	> 125	1.6	77282	16060	No[1]	Yes

[1] Vortex SERIES [9] was used in place of NEWA to define atmospheric stability.

DNV pre and post construction EPA analyses were completed for each wind farm, defining the wind farm configurations considered. The SCADA power data was cleaned, turbine-by-turbine, to remove any values where turbines were not available, underperforming, or in an unknown state. On sites L2, XL1, and XL2, where ambient wind speed gradients were considered within the pre-construction EPA we also considered these within the analysis.

These data sets have not been used to develop or 'train' the turbine interaction models used in this study, nor do they include the Horns Rev I or Nysted offshore wind farms.

2.2. *Derivation of time series flow cases*

The atmospheric conditions at a reference location are defined for each time step in the analysis as follows to form a time series of flow cases we will evaluate in each model.

2.2.1. Derivation of reference wind speeds. The models considered use a time series of wake-free reference wind speeds as input. This time series can be defined at a single reference location, i.e., a virtual reference mast U_{ref}^{mast}, and at each turbine i, U_{ref}^{i}. Where a wind flow model is available it is used to convert U_{ref}^{mast} into U_{ref}^{i} via:

$$U_{ref}^{i} = U_{ref}^{mast} * S_{mast}^{i}(D) \quad (1)$$

where S_{i}^{mast} is the flow model speedup from the mast to the turbine for the wind direction, D. For the 3 offshore wind farms in this study with no wind flow model defined, we assume $U_{ref}^{i} = U_{ref}^{mast}$.

Firstly, we select front-row turbines. Wake free turbines are first selected for 5-degree wind direction bins for each wind farm. To aid with the selection of the front-row turbines the WindFarmer Park model was run at 1 degree direction resolution to predict waked wind speeds at all turbines in the farm. These speeds were compared to free stream conditions and candidate front row turbines were suggested where free stream wind speeds persist in the waked wind speed results. The suggested front row turbines were reviewed conservatively to discard turbines either close to being within a wake, or any turbines not on the perimeter of the wind farm. We wish to avoid consideration of turbines where flow accelerations may be expected between upstream turbines.

For each time step considered, the air density corrected power curve (following the IEC 61400-12-1 correction method) is used to obtain the wind speeds local to each of wind farm turbines from the SCADA data, U_{SCADA}^{i}. To define U_{ref}^{mast} for a time step we take the mean of wind speeds calculated at the front row turbines after extrapolation to the reference location by application of the flow model:

$$U_{ref}^{mast} = \frac{1}{N}\frac{\Sigma_{i=1}^{N} U_{SCADA}^{i}}{S_{mast}^{i}(D)} \quad (2)$$

Where the turbines i are those considered in the front row for the wind direction D.

The U_{SCADA}^{i} for front row turbines differ due to both ambient wind speed variations and turbine interactions. By taking the mean across many front row turbines we can mitigate some biasing impacts from turbine interactions relationships on the definition of the reference wind speed, however this is a noted source of uncertainty.

2.2.2. Air density time series. An air density time series is defined from reanalysis time series data from the Modern-Era Retrospective analysis for Research and Applications (MERRA-2) [10] for each site. This time series is scaled to a long term mean air density at hub height defined during the pre-construction EPA for each wind farm. The scaled air density time series is up-sampled to match the 10-minute SCADA data for use as a model input and for reference wind speed derivation.

2.2.3. Atmospheric stability time series data. Atmospheric Stability is represented in our analysis with the Monin-Obukhov length (MOL). The Inverse Obukhov Length (RMOL) variable from the

NEWA [8] was used for wind farms where NEWA could provide data concurrent with the wind farm SCADA data to define a time series of MOL. Extreme MOL values were controlled by capping the data within the following range, following the definition of Gryning [11]:

$$\begin{aligned} &\text{For MOL} > 0: && 10 <= \text{MOL} <= 2000 \\ &\text{For MOL} < 0: && -2000 <= \text{MOL} <= -50 \end{aligned} \quad (3)$$

2.2.4. Wind speed standard deviation. Where NEWA could provide time series data concurrent with the wind farm SCADA data, the turbulent kinetic energy, TKE, variable is used to define a time series of wind speed standard deviation, σ, via $\sigma = \sqrt{TKE}$. We also consider WTI long-term statistics from the relevant DNV EPA to test model sensitivities to this input. WTI posits a turbulence characteristic that varies with 10-minute mean wind speed but is otherwise not time varying. We rely solely on WTI statistics at L2 and XL2 where the SCADA data was not concurrent with NEWA.

2.3. Data filtering

The methodology described here to use wind turbines as wind speed measurement devices has inherent uncertainty, so filters are applied to both measured and modelled power time series to improve confidence in our comparisons. The filters are summarised in **Table 2** and visualised at one site in **Figure 1**.

Table 2. Filters applied to validation power time series

Filter	**Condition / approach**
Availability	Remove record when > 5% turbines are unavailable, derated or in an unknown state.[a]
Noisy records	Remove records where $\sigma_{P_{\text{measured}}} > \frac{\bar{P}_{\text{measured}}}{2}$
Power curve "knee"	Remove records where > than 1% of turbines are at > 90% of rated power.
Near low wind speed cut-in	Remove records where $U_{\text{ref}} < 4$ m/s

[a] At smaller wind farms S1 and M1 we require 100% availability

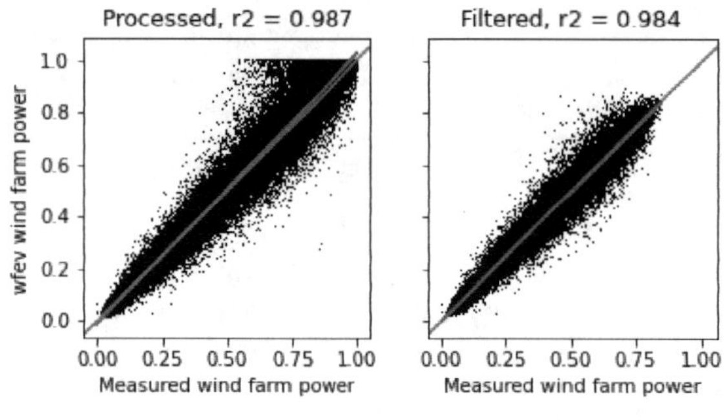

Figure 1. Scatter plot comparison of wfEV (3.2.) modelled and measured wind farm power time series, before and after application of filters at site S1. The power time series is normalised by the total wind farm capacity.

In section 2.1. we noted how the SCADA power data were cleaned. For time stamps and turbines where cleaning has been performed on specific measured turbine powers, we also remove the corresponding turbine power from our modelled power predictions. If the proportion of turbines unavailable were too large for a timestep, the operational state of the wind farm will be poorly represented in the model and the model will predict extra turbine interaction losses. We therefore exclude records where fewer than 95% of turbine records remain after cleaning.

The cleaned measured power data contains further noise uncorrelated with model predictions. This could possibly be attributed to missed cleaning or other factors such as significant changes in wind speed and direction across the wind farm within a 10-minute record. A simple filter is applied where we drop records where the standard deviation of measured turbine power is greater than 50% of the mean turbine power for the record.

The Power Curve Working Group, https://pcwg.org/, has outlined how turbine power curves become unrepresentative of actual turbine performance in conditions where atmospheric turbulence is in an outer range compared to that for which the power curve was defined [12]. Turbine performance deviations from the power curve are greater for low wind speeds and those where power approaches rated power. Furthermore, the curvature of the power curve is greatest just above the low wind speed cut-in and approaching rated power at the "knee" of the power curve, so in these regions any uncertainty in power results in a correspondingly greater uncertainty in derived U_{ref}. To control these uncertainties, we filter out cases with $U_{\text{ref}} < 4\ \text{m/s}$. We also exclude records where more than 1% of turbines are above 90% of rated power.

2.4. Comparison metrics

Modelled and measured power are compared on a time series point by point basis rather than on their annualised means. The main indicators used on farm and turbine level are as follows:

Bias in power output for each time stamp
$$\Delta P = P_{\text{modelled}} - P_{\text{measured}} \tag{4}$$

Mean bias over time stamps N
$$\text{MBE}_{\text{P}} = \frac{1}{N}\sum_{i=1}^{n} \Delta P_i \tag{5}$$

Relative mean bias over time stamps N
$$\text{rMBE}_{\text{P}} = \frac{MBE_P}{\bar{P}} \text{ with } \bar{P} = \frac{1}{N}\textstyle\sum_{i=1}^{n} P_{\text{measured}\,i} \tag{6}$$

Wake loss (modelled and measured)
$$\text{Wake Loss } i = 1 - \frac{P_i}{P_{\text{ref}}^i} \tag{7}$$

To calculate the wake loss, (7), P_{ref}^i is derived from the U_{ref}^i for every turbine. Where a wind flow model is available, P_{ref}^i includes the power variations due to ambient wind speed gradients, but not those relating to turbine interactions. We refrain from calling P_{ref}^i the gross turbine power output, as it does not include the overall wind farm blockage correction.

Whilst the farm level errors MBE_{P}, help us understand overall bias in the model, we present distributions of ΔP at a turbine level. We hypothesise that models with a narrower distribution of time-series turbine ΔP will produce better model results when considering new wind farms with different geometries and more turbines than those in the validation: they are less likely to rely on positive and negative biases cancelling one another.

By deriving ΔP time series for all turbines we have a rich data set that can be sliced to understand modelling error sensitivity to various factors such as atmospheric stability, wind speed, wind direction and location within farm.

3. Models considered

Four models are considered in this study with differing inputs required and levels of complexity. The inputs required to define a 10-minute flow case for each model are summarised in **Table 3**.

Table 3. Models considered and their dependencies

Model	Dependencies				
	Wind direction	Wind speed	Wind speed standard deviation	Air density	Monin-Obukhov Length
WindFarmer PARK (wfPARK)	X	X		X	
WindFarmer Eddy Viscosity (wfEV)	X	X	X	X	
Stratified Eddy Viscosity Model (sEVM)	X	X	X	X	X
CFD.ML[d]	X	X		X	

[a] PARK model with large wind farm correction
[b] Ainslie model with large wind farm correction
[c] DNV model for wind farm control modelling
[d] DNV neural network model trained on CFD simulations

3.1. CFD.ML

CFD.ML predicts turbine interaction loss using a graph neural network model that has been trained to replicate results from Computational Fluid Dynamics (CFD) simulations. The network in this study has been trained on the results from over 350 steady-state Reynolds-Averaged Navier-Stokes (RANS) simulations run on 25 different offshore wind farms. The training set includes a range of wind direction and wind speeds, but the parameters used to define the potential temperature profile are largely the same throughout the set. In general terms these parameters specify a neutral boundary layer with an overlying capping inversion and stably stratified free atmosphere. CFD.ML is similar to the model described in [13], but has been trained on different data and extended to handle the full range of turbine operating wind speeds. This surrogate model is unique within this study as it accounts for the coupled phenomena of blockage and wake effects together while still producing results as fast or faster than the other models. Note that for two of the wind farms in this validation, a differently trained CFD.ML model was used where the only change was removal of RANS results of the two wind farms from the training set. In other words, all the CFD.ML predictions in this validation are made for wind farms the model has never seen before.

It is worth noting that because CFD.ML predicts wakes and blockage together, the wind speed inputs required for CFD.ML are free stream wind speeds. Within the CFD.ML simulations in this study, the blockage impacted U_{ref} wind speeds at the front row turbines are first used to derive an estimate of the free stream wind speeds. To predict turbine interaction impacted incident wind speeds, U_i, at all turbines CFD.ML is then run using its own derived estimate of the free stream wind speed.

3.2. WindFarmer Eddy Viscosity Model

The WindFarmer Eddy Viscosity (wfEV) wake model is based on the work of Ainslie described in [14] [15]. The wfEV wake has a Gaussian wind speed deficit profile perpendicular to the wind direction, defined by a centre-line-deficit. The development of the velocity deficit field is defined by solving the Reynolds-Averaged-Navier-Stokes equations in two dimensions using axisymmetric co-ordinates. The Eddy Viscosity model automatically observes the conservation of mass and momentum in the wake. An eddy viscosity, averaged across each downstream wake section, is used to relate the shear stress to gradients of velocity deficit. The mean wind speed field can then be obtained by a linear superposition of the wake deficit field and the incident wind flow field

Two wfEV models are used in this study, wfEV (newa) and WFev (wti), corresponding to the two ambient turbulence assumptions noted in 2.2.4.

3.2.1. Wake added turbulence intensity. The wfEV model determines the incident turbulence intensity at each turbine through a summation of a calculated wake added turbulence and the ambient turbulence for the specific location and flow conditions. For a turbine in the free wind stream, the incident turbulence is simply the ambient turbulence level. For a turbine within a wind farm, it is necessary to calculate the wake added turbulence level that results from the presence of upstream turbines.

Wake added turbulence levels are calculated using an empirical characterisation developed by Quarton and Ainslie [16]. This characterisation enables the added turbulence in the wake (I_{add}) to be defined as a function of ambient turbulence (I_{amb}), the turbine thrust coefficient (C_t), the distance downstream from the rotor plane (x) and the length of the near wake. The characterisation used in wfEV was subsequently amended slightly to improve the prediction:

$$I_{add} = 5.7C_t^{0.7}I_{\text{amb}}^{0.68}\left(\frac{x}{x_n}\right)^{0.6}$$

where x_n is the calculated length of the near wake.

It is worth noting that where I_{add} is greater, the rate wake recovery rate increases. In this way the wfEV model considers how wake added turbulence impacts the rate of wake dissipation throughout the wake's development.

3.2.2. Dominant wake superposition method. The WindFarmer eddy viscosity wake model assumes the dominant wake from upstream turbines persists when 2 or more wakes overlap. This means the strongest wake deficit or greatest wake added turbulence is considered when integrating wakes over a wake impacted downstream turbine rotor to define the incident wind speed.

3.2.3. Large wind farm model. It is well-recognised that wind turbines do not only react passively to the wind regime, but at the same time they are part of it [17], [18], [19], [20]. Qualitatively, the wind farm can be thought of as an area of higher roughness where the boundary layer profile is disturbed by the extraction of momentum by the wind turbines. Offshore wind farms cause a more pronounced local effect than onshore wind farms, due to the lower roughness offshore, and this impact has been likened to an onshore forest.

A large wind farm (LWF) model was developed by DNV [3] and [2], based on modelling the disturbance caused by each individual turbine with an empirically defined exponential recovery profile defined in [21]. The LWF model provides two main tunable parameters: a base and an increased roughness. In this study the base roughness was set as 0.0002 and the increased roughness as 0.0192. At the time of writing, these are the values used in DNV Offshore EPAs.

This LWF model is applied as standard when using the wfEV model to account for the stronger turbine interactions present deep in larger wind farm arrays and is applicable to onshore projects [22] as well as offshore projects. Alternative roughness settings with a greater increased roughness setting have been shown to yield improved results in stable atmospheric conditions for onshore projects [23].

It is worth noting that the LWF model is specific to the choice of the dominant wake superposition method. If an alternative wake superposition method were chosen, then a new approach would be required to model conditions deep in a wind farm array.

3.2.4. Blockage correction. When using the wakes-only wfEV model in an EPA, DNV would apply an additional blockage correction loss factor. As previously noted, the magnitude of this correction cannot be estimated with the methodology presented here so it is not considered within this study.

3.3. Stratified Eddy Viscosity Model

The Stratified Eddy Viscosity Model (sEVM), described thoroughly in [1], is an engineering wake model based on the Ainslie model [24], [15], like the wfEV model. It includes a stability correction in the eddy diffusivity of momentum term, also the assumption that Hub Height ≈ Rotor Diameter is removed. In [1] and [25] these modifications were demonstrated to yield better results than the standard wfEV model in predominantly stable, and predominantly unstable conditions.

A key motivation for development of the sEVM is to support the dynamic simulations required for design and evaluation of wind farm control strategies where active yaw control is used for wake steering to reduce impacts on downstream turbines, aiming to boost wind farm power outputs and control turbine loading [26], [27]. Within the context of this study, we use the sEVM to model steady state conditions, i.e., 10-minute average flow conditions, to explore how the model performs in the context of the EPA.

3.4. WindFarmer Park Model.

The WindFarmer Park Model (wfPARK) is a simple model of the wake that is based on the same algorithm used in PARK [28] [29]. This two-dimensional model uses the momentum theory to predict the initial profile from the thrust coefficient assuming a rectangular wind speed profile, and that the wake expands linearly behind the rotor.

The downstream wind speed is calculated using the following equation:

$$U_w = U_i\left[1-\left(1-\sqrt{1-C_t}\right)\left(\frac{D}{D+2kx}\right)^2\right] \tag{8}$$

Here Ui is the axial wind speed incident on the turbine, Ct is the thrust coefficient and k is the wake decay constant. The parameter k used in this study was set to a fixed value of 0.04.

The same LWF model and dominant wake superposition approaches, described for the wfEV model above, are applied to the wfPARK model.

3.5. Power prediction

Each model described above predicts wind speeds that must be converted into power prior to validation. We use the power curve considered for each wind farm in the DNV pre-construction EPA, with the air density correction for pitch regulated turbines from IEC 61400-12. The correction is applied to the wind speeds of the turbine power curve for each time step in the analysis using the air density for that time step according to the equation:

$$u_{\text{local}} = u_{\text{source}}\left(\frac{\rho_{\text{src}}}{\rho_{\text{local}}}\right)^{\frac{1}{3}} \tag{9}$$

where u_{local} is the local air density wind speed and u_{source} is the power curve wind speed. Turbine power for a time step is then derived through linear interpolation of the power curve's corrected speeds, u_{local}, at the turbine incident wind speed, U_i.

4. Results

4.1. Wind farm mean bias errors

The overall wind farm power relative mean bias error ($rMBE_P$), as defined in 2.4., is presented for each wind farm in **Table 4**. These statistics represent the modelling error for the entire wind farm power, allowing under and over-predictions at turbines and for specific conditions to cancel.

Table 4. Wind farm power rMBE [%] for each model and wind farm. The result closest to zero is displayed in bold type. Positive values imply overprediction of wind farm active power.

$rMBE_P$ [%]		Wind Farm					
		S1	M1	L1	L2	XL1	XL2
Model	**CFD.ML**	-0.7	-0.9	-2.0	-2.0	-0.7	3.0
	SEVM	**0.0**	**0.0**	-1.9	**-0.5**	**0.5**	**0.8**
	wfEV (newa)	-0.8	-1.9	**-0.9**	-	1.0	-
	wfEV (wti)	-0.9	-1.8	-1.7	-0.9	-1.3	1.3
	wfPARK	-2.5	-5.2	-7.3	-2.9	-7.4	2.2

4.2. Turbine rMBE distributions

When looking at wind farm level errors only, a particular model could appear to perform well but exhibit large modelling errors when we look more closely at errors for specific turbines, or for specific atmospheric conditions. To explore the modelling errors at each turbine we present a visualisation of rMBE at each turbine, as a frequency density distribution. These graphics may be thought of as a frequency distribution of "pattern of production" modelling errors. A narrower peak indicates a better prediction of relative pattern of production trends, and a peak closer to zero indicates better prediction of wind farm $rMBE_P$. Turbine $rMBE_P$ distributions are shown for every model and wind farm in **Figure 2**.

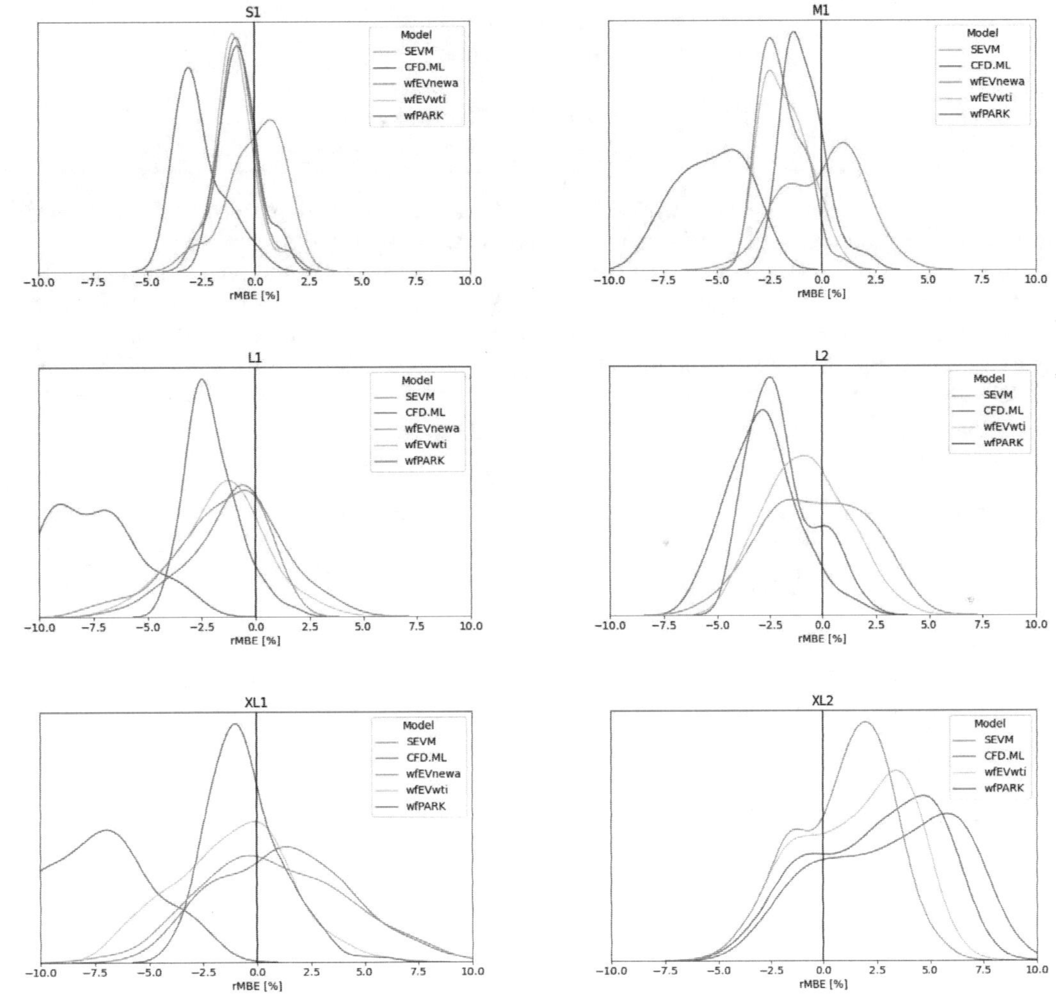

Figure 2. Frequency density distributions of turbine power rMBE [%] representing the pattern of production modelling errors. Each model variant is compared for each wind farm. Each curve integrates to have an area of 1, so the density axis is omitted for readability.

4.2.1. Error trends with wind farm scale. $rMBE_P$ is shown to exhibit some dependency with respect to the wind farm scale. For all models we see the smallest modelling errors at the smallest wind farm, S1. The spread of turbine $rMBE_P$ seen in Figure 2 increases as we move from the smallest wind farm, S1 to larger farms, with the greatest spread of turbine $rMBE_P$ at the XL farms. We also see a notable shift towards a more positive $rMBE_P$ at the larger XL wind farms for the CFD.ML, SEVM and wfEV models. This indicates that, on average, our models underpredict wake losses at these XL farms.

In addition, wind farm XL2 predictions exhibit the greatest spread of turbine $rMBE_p$ results. The family of bi-modal error distributions seen for all models at XL2 warrants further investigation to understand the drivers of these errors. We suggest that further investigation of the deep array effects measured within the largest wind farms should be informative for further model refinement.

4.2.2. Model responses to atmospheric stability. The modelling results were sliced in to 3 stability classes defined using the criteria on Monin-Obukhov Length (MOL) in **Table 5** and [11].

Table 5. Stability classifications

Stability class	Monin-Obukhov Length ranges
Unstable	$-500 \leq$ MOL < -50
Neutral	MOL > 500 or MOL < -500
Stable	$10 <$ MOL ≤ 500

The error responses to atmospheric stability were then examined for each site and model. In **Figure 3** an examination of the wfEV modelling errors shows that a time series wind speed standard deviation input from NEWA yields significantly lower errors than the turbulence intensity-by-speed WTI input for the stable and unstable classes. With WTI inputs, in stable conditions the rMBE peak has significant positive bias (wakes are underpredicted), and in unstable conditions the peak is negatively biased (wakes are overpredicted). Similar observations were made for wfEV results for sites S1, M1 and L1. However this pattern is not present for XL1 for which stable case power is overpredicted with both WTI and NEWA turbulence inputs.

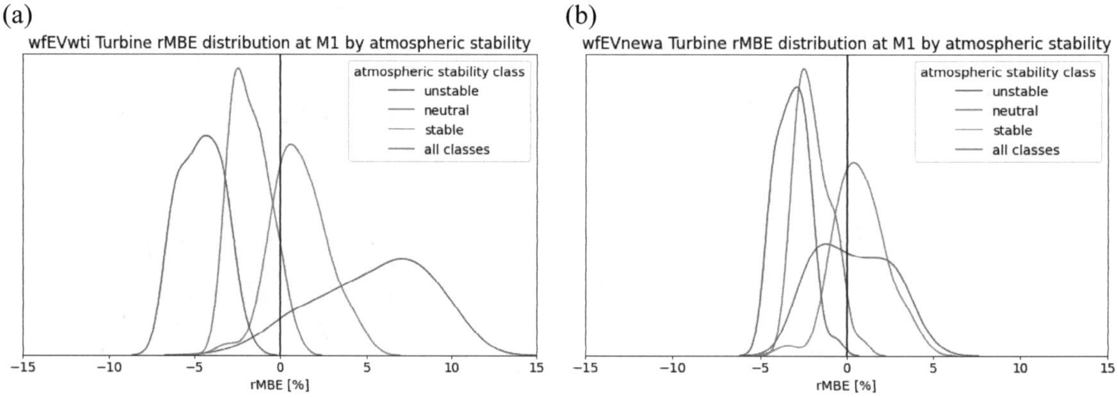

Figure 3. wfEV turbine MBE_P distribution split by atmospheric stability at wind farm M1 with wind speed standard deviation inputs defined by (a) "WTI" long-term statistics only dependent on wind speed and, (b) "NEWA" time series.

In **Figure 4**. the SEVM model is shown to underpredict power (over-predict wakes) in stable conditions. Similar behavior was seen at all sites except XL2. It is possible that the SEVM is oversensitive to atmospheric stability in cases where multiple wakes have been combined; a hypothesis to test against further data sets.

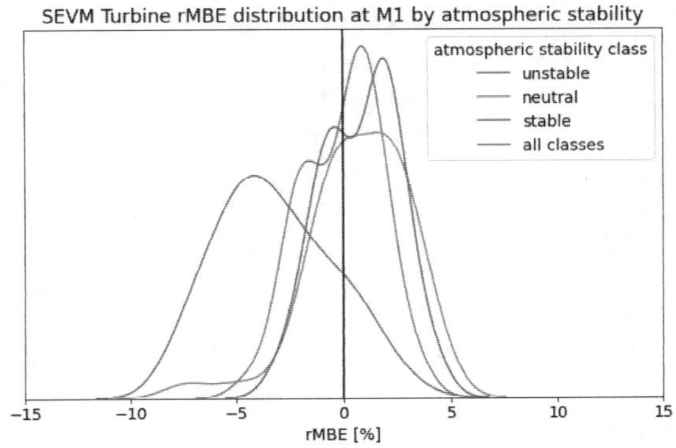

Figure 4. SEVM turbine $rMBE_P$ split by atmospheric stability class at wind farm M1.

5. Discussion

5.1. Modelling performance.

The range and types of sites analysed has allowed us to show some of the ways in which modelling uncertainties can vary across farm scales, model types and stability classes but given the limited number of sites considered it is not appropriate to further quantify these uncertainties. Instead we provide a qualitative discussion of modelling performance intended to guide future studies.

5.1.1. wfPARK model performance. The wfPARK model exhibits the greatest range and magnitude of errors, both when looking at either the wind farm power or turbine power. It is noted that the wake expansion constant set to k=0.04 could be tuned to bring the average wind farm rMBE closer to zero, or k could be made to vary with other input parameters. We should also note that any tuning of wfPARK must be specific to the implementation of PARK described in section 3.4. considering the wake superposition and LWF models described there. The PARK model is highly dependent on its parameter tuning, so will have increased uncertainty when applied in scenarios dissimilar to the training data. DNV has therefore not pursued further development of wfPARK.

5.1.2. wfEV model performance. The wfEV model errors exhibit dependency on the choice of wind speed standard deviation input, either NEWA time series or WTI long-term statistics. When looking at overall results at a wind farm level in **Table 4** or turbine level in **Figure 2**, the differences in wfEV modelling errors with WTI or NEWA inputs are minor, but the differences become remarkable when results are split by atmospheric stability classes. Turbulence alone is not enough to model the impact of atmospheric stability on turbine interactions [1], [25], but lower turbulence intensity is more common in stable atmospheric conditions. The WTI definition of turbulence is an average for all atmospheric conditions, so for specific stable, low turbulence cases the wfEV modelled wakes with the WTI input will recover too quickly. It is therefore not surprising to find that where atmospheric stability is strongly time varying (as opposed to varying only on mean wind speed) that we see the NEWA fed model yielding significantly reduced modelling uncertainties comparted with WTI.

These considerations illustrate the importance of selecting good wind speed standard deviation input data for Ainslie based models, and they will also apply to the SEVM given its similar basis.

5.1.3. CFD.ML modelling performance. For all wind farms except XL2 the pattern of production distribution for CFD.ML is narrower than that seen for the other models. We hypothesise that this is because whilst all the models have a capacity to model wake effects, it is only CFD.ML that can also model the variation in turbine interaction effects across the front row of a wind farm.

The CFD.ML model considered here cannot consider the impact of atmospheric stability or turbulence on turbine interactions, however the CFD.ML training could be extended to consider these factors, given appropriate further RANS simulations.

Looking at wind farm level $rMBE_P$ in **Table 4** we see CFD.ML perform comparably to the wfEV and SEVM models, except at XL2. CFD.ML can be can also be expected to have elevated uncertainty for wind farms with configurations that differ strongly from its training data. A possible near-term application of CFD.ML is to combine it with other models to improve pattern of production predictions, or to use it to interpolate results from computationally expensive RANS simulations. CFD.ML is a promising approach to fast turbine interaction modelling, and with additional training on more farms and conditions, it is likely that errors can be reduced.

5.1.4. SEVM modelling performance. The SEVM model performs best at 4 out of 6 sites when viewing the wind farm level rMBE statistic. When viewing the turbine MBE distributions in **Figure 2**, we see some bi-modal distributions, which are potentially caused by the over-sensitivity to atmospheric stability demonstrated in **Figure 4**. This was also observed for one example within [7]. It is noted that the SEVM does not use the same wake superposition or LWF model as wfEV or wfPARK and there remains significant scope for experimentation with alternative wake combination approaches in the SEVM.

5.2. Dependency of MBE on wind speed.
In section 2.3. we discussed concerns regarding the accuracy of our derived reference wind speeds at wind speeds near turbine cut-in and approaching rated power. To explore the impacts of this concern the turbine $rMBE_P$ distributions were split by wind speed, as shown in **Figure 5a** for site S1. A significant bias towards model underprediction at low wind speeds is observed, and an overprediction at high wind speeds. However, when we look at the un-normalised MBE_P in kW in **Figure 5b** the central estimates of bias from the lower speeds align with those in the mid-range of the power curve: the impact on overall results from errors at low wind speeds is lower due to the lower power predictions at these speeds. We perhaps should be more concerned about overprediction errors in the high wind speed cases.

We expect the error dependency on wind speed to be an artifact of the validation methodology rather than the turbine interaction models. Future work could benefit from a consideration of extra factors in the reference wind speed prediction, possibly including extra SCADA signals such as rotor speed and pitch angle. Power curve corrections for turbulence intensity should also be considered. Further validation of the derived U_{ref} wind speeds against measurements could also build confidence in our absolute wind farm level error metrics. In this study these biases were largely mitigated by discarding wind speeds < 4 m/s and cases with measured power > 90 % of rated power.

(a) (b)

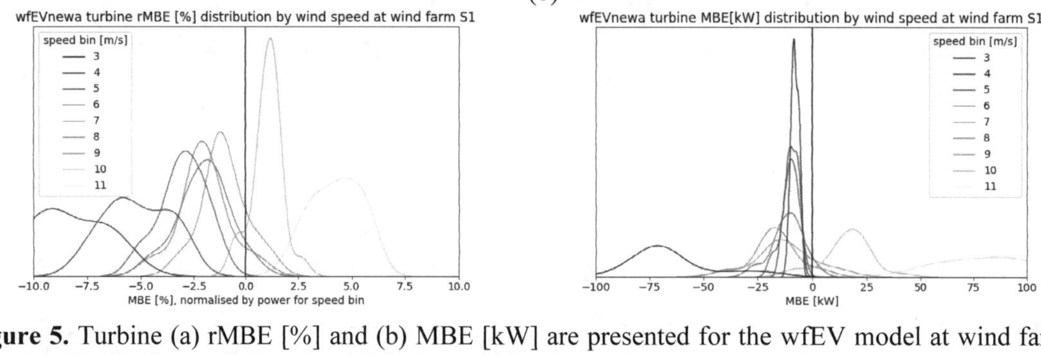

Figure 5. Turbine (a) rMBE [%] and (b) MBE [kW] are presented for the wfEV model at wind farm S1, with distributions split into 1 m/s wind speed bins. In (a) the rMBE at each turbine first normalised to the mean power for that turbine in that wind speed.

6. Conclusions

We have presented a validation approach based on power time series data from the turbine SCADA system and used it to evaluate the performance of four wind farm models across a range of wind farm scales from <25 turbines to >125 turbines. The spread of turbine power modelling errors was seen to increase for larger wind farms. For the two "extra-large" farms considered we saw the models overpredict power and underpredict wakes, suggesting further investigation of wake modelling in the deep array is worthwhile for the largest offshore wind farms.

The validation methodology was found to have wind speed dependent biases at wind speeds near turbine low speed cut-in and rated power. Filters were applied to mitigate impacts of these biases on the analysis, but they reduced data available from which we can draw conclusions on the wind farm level mean bias error statistics. Future refinement of the wind speed prediction method may mitigate some of these concerns and allow filters to be relaxed, improving data recovery, and strengthening of conclusions.

The spread of turbine level mean bias error in predictions of power from each model were explored to compare models. We found wfPARK to be the least reliable model in the set considered. wfEV, SEVM and CFD.ML exhibited more comparable performance. CFD.ML was found to replicate the wind farm pattern of production remarkably well given its novel approach to turbine interaction modelling.

The SEVM model, an evolution of wfEV considering atmospheric stability, yields comparable and in some cases improved, modelling results compared with wfEV at a wind farm level. However, the version the SEVM tested here has been shown to be overly sensitive to atmospheric stability, predicting too strong turbine interactions in stable conditions for larger wind farms.

The importance of appropriate turbulence intensity inputs has been demonstrated for the wfEV model through an examination of the results sliced by atmospheric stability. Improved accuracy was achieved by considering time series turbulence inputs rather than defining turbulence intensity purely as a function of wind speed.

This study described the basis for an automated and generalised approach for validating and comparing turbine interaction models. Using DNV's access to quality-controlled SCADA data we now look forwards to refining our method, expanding our validation database, and testing more models in future.

7. References

[1] Ruisi R and Bossanyi E, "Engineering models for turbine wake velocity deficit and wake deflection. A new proposed approach for onshore and offshore applications," in *J. Phys. Conf. Ser.*, May 2019.

[2] Schlez W and Neubert A 2009 New Developments in Large Wind Farm Modelling *EWEC*, (Marseille, France)

[3] Schlez W, Neubert A and Smith G M, 2006 New Developments in Precision Wake Modelling *DEWEK*, (Bremen)

[4] Bleeg J, Purcell M, Ruisi R and Traiger E 2018 Wind Farm Blockage and the Consequences of Neglecting Its Impact on Energy Production *Energies* **11** 1609.

[5] Hansen K S, Barthelmie R J, Jensen L E and Sommer A 2012 The impact of turbulence intensity and atmospheric stability on power deficits due to wind turbine wakes at Horns Rev wind farm. *Wind Energy* **15** p.183-196.

[6] Peña A and Rathmann O 2014 Atmospheric stability-dependent infinite wind-farm models and the wake decay coefficient. *Wind Energy* **17**(8) p. 1269–1285

[7] Neubert A and Downes P 2021 Validating time-domain energy assessments: wake modelling

considerations and recommendations *WindEurope Technical Workshop*

[8] Dörenkämper M et. al. 2020 The Making of the New European Wind Atlas – Part 2: Production and evaluation *Geoscientific Model Development* **13**(10) pp. 5079--5102

[9] Vortex SERIES, https://vortexfdc.com/wind-speed-time-series/

[10] Rienecker M, et. al. 2011 MERRA: NASA's Modern-Era Retrospective Analysis for Research and Applications *vol. J. Climate* **24** pp 3624-3648

[11] Gryning S E, Batchvarova E, Brümmer B, Jørgensen H and Larsen S 2007 On the extension of the wind profile over homogeneous terrain beyond the surface boundary layer *Boundary-Layer Meteorology* **124** pp. 251-268

[12] Lee J C Y, Stuart P, Clifton A, Fields M J, Perr-Sauer J, Williams L, Cameron L, Geer T and Housley P 2020 The Power Curve Working Group's assessment of wind turbine power performance prediction methods *Wind Energ. Sci* **5** p 199–223

[13] Bleeg J 2020 A Graph Neutral Network Surrogate Model for the Prediction of Turbine Interaction Loss *J. Phys.: Conf. Ser.* **1618** 062054.

[14] Ainslie J F, 1985 Development of an eddy viscosity model for wind turbine wakes *British Wind Energy Association conference* (Oxford)

[15] Ainslie J F 1988 Calculating the flowfield in the wake of wind turbine *Journal of Wind Engineering and Industrial Aerodynamics* **27**

[16] Quarton D and Ainslie J F 1990 Turbulence in wind turbine wakes *J. Wind Eng.* **14** pp. 15-23

[17] Bossanyi E, Maclean C, Whittle G, Dunn P, Lipman N, Musgrove P 1980 The Efficiency of Wind Turbine Clusters *International Symposium on Wind Energy Systems* (Lyngby, Denmark)

[18] Crafoord C 1975 An Estimate of the Interaction of a Limited Array of Windmills (Department of Meteorology, University of Stockholm, Sweden, Report DM-16.)

[19] Milborrow D J 1980 The Performance of Arrays of Wind Turbines *Journal of Industrial Aerodynamics* **5** pp 403-430

[20] Newman B G 1977 The Spacing of Wind Turbines in Large Arrays *Energy Conversion* **16** pp 169-171

[21] Smith G, Neubert A and Schlez W 2011 Impact of large neighbouring wind farms on energy yield of offshore wind farms in *EWEA Offshore Conference* (Amsterdam) 2011.

[22] Johnson C, Graves A M, Tindal A, Cox S, Schlez W and Neubert A 2009 New developments in wake models for large wind farms. *AWEA WINDPOWER Conference* (Chicago, USA)

[23] Geer T 2017 Improving confidence in wake predictions through operational validations *Wind Europe Offshore* (London)

[24] Ainslie J F 1983 Development of an Eddy Viscosity model of a Wind Turbine Wake *CERL Memorandum* TPRD/L/AP/0081/M83

[25] Downes P, Harrison M, Bossanyi E, Ruisi R and Skeen N 2021 Validating wake deflection and atmospheric stability's impact on wake deficit for wind farm control *Wind Europe Electric City* (Copenhagen)

[26] Bossanyi E and Ruisi R 2021 Axial induction controller field test at Sedini wind farm *Wind Energy Science* **6** pp. 389-408

[27] Natarajan A et. al. 2021 www.totalcontrolproject.eu TotalControl Deliverable D2.3

[28] Jensen N 1983 A note on wind generator interaction *RISØ DTU* **M-2411**, 1983.

[29] Katic I, Høstrup J and Jensen N 1986 A simple model for cluster efficiency *EWEC*

Acknowledgments
B Williams, T Mercer, D Rex, M Chequer, M Harrison

Journal of Physics: Conference Series 2257 (2022) 012011 doi:10.1088/1742-6596/2257/1/012011

Assessment of Vortex Induced Vibrations on wind turbines

Dimitris I. Manolas[1, 2], Panagiotis K. Chaviaropoulos[1], Vasilis A. Riziotis[2]

[1]iWind Renewables, PC, Kleisthenous 166 Gerakas, GR15344 Athens, Greece
[2]School of Mechanical Engineering, National Technical University of Athens, Heroon Polytechniou 9 Zografou, GR 15780, Athens, Greece

corresponding author's e-mail: d.manolas@iwind.gr

Abstract. Modern wind turbines are prone to Vortex Induced Vibrations (VIV). In the present work, an engineering semi-empirical framework is proposed that assesses VIV aero-elastic instabilities of wind turbine configurations. The procedure employs engineering tools relying on airfoil polars. It uses the state-of-the-art aero-elastic tool hGAST along with the EUROCODE VIV framework for steel structures extended to wind turbine configurations. The aero-elastic tool provides the missing modal input data (i.e. modal frequencies, total structural plus aerodynamic modal damping and modeshapes) to evaluate the semi-analytical expressions of the displacement and load amplitudes. Numerical results for single- and two-bladed configurations of the NREL 5MW Reference Wind Turbine (RWT) during assembly are presented, assessing turbine loads under the most unfavourable VIV scenarios examined.

1. Introduction

As the wind turbines are getting bigger in size, designs are becoming lightweight and consequently the ratio of the aerodynamic over the inertial loading increases. Advanced blade manufacturing processes increase the accuracy of construction and bill of materials, while they reduce blade's structural damping, which is further compromised by the more extended use of carbon composites. All the above, along with the increasing flexibility of modern designs that strengthen aero-elastic couplings and lower system natural frequencies, make wind turbines prone to Stall and Vortex Induced Vibrations (VIV). VIV is the aero-elastic instability a structure undergoes when the cross flow oscillating lift force, due to shedding of vortices, resonates with one of the system's natural frequencies. Severe VIV occur when the shed vortices form a coherent wake structure, which in wind turbines requires the rotor either locked or in idling mode. On the contrary, VIV are suppressed when the coherent structure of the shed vortices is destroyed, preventing vortex lock-on. Therefore, VIV rarely occur during normal operation but may appear under parked/idling conditions or during maintenance and assembly procedures. The question that needs to be answered by the manufacturers is whether the wind turbine is safe against VIV.

While the literature regarding VIV of cylindrical structures is rich, relevant experimental [1] or computational [2]-[6] works on wind turbine blades are scarce, attracting high attention the last decade. For a recent, detailed literature survey on VIV the reader is cited to [6]. Below we make reference to VIV studies on wind turbines (WTs). In early works VIV was studied on airfoil sections [2]-[4], while recent works by DTU consider the elastic blade clamped at its root [5]-[6]. In all the above references, free wake [2] or CFD [3]-[6] advanced aerodynamic methods were employed, primarily focusing on the aerodynamics of the complex VIV phenomenon either in stiff or moving/elastic configurations. However, due to the complexity of the problem and of the high computational cost, none of the works assessed VIV of the entire (coupled) WT configuration.

In this work we present an engineering framework for performing VIV analysis of wind turbine configurations based on EUROCODE "VIV Approach 1" provisions [7]. This framework is incorporated in the aero-elastic tool hGAST [8]-[10] and relies on tabulated airfoil data (i.e. polars). Numerical results for single and two-bladed configurations of the NREL 5MW Reference Wind Turbine (RWT) [11] during assembly are presented, assessing turbine loads under the most unfavourable VIV scenarios examined. The paper is structured as follows; in section 2 the semi-empirical framework for VIV analysis is presented, in section 3 indicative numerical results are provided, while section 4 concludes the present work.

The contribution of the present work is the formulation of an engineering approach for wind turbines that allows performing VIV analysis of the entire (coupled) WT configuration. The engineering knowledge is combined with a state-of-the-art aeroelastic tool to assess turbine loads under the most unfavourable VIV scenarios, addressing excitation due to vortex shedding from the rotor or the tower at blade or tower frequencies. The established framework applies to all low damped modeshapes that may suffer from VIV, including 2nd tower modes in cases of tall towers with relatively low second tower frequency.

2. Methodology

The engineering semi-empirical framework for VIV analysis we establish in our present work is based on EUROCODE "Approach 1" [7] framework for steel structures, properly extended to wind turbine configurations including tower, nacelle, and rotor arrangements. It is incorporated in the non-linear, multi-body aeroelastic tool hGAST [8]-[10], in which both time-domain and eigenvalue stability analyses are available. In time-domain, hGAST applies the EUROCODE load as dynamic external force, while modal analysis provides all additional information needed (i.e. mode shapes, frequencies and total damping) to evaluate the semi-analytical expression of the displacement and load amplitudes. In the following sections we provide the background of EUROCODE approach, along with modelling assumptions and extensions made by the present work.

2.1. *Baseline EUROCODE "VIV approach 1" for steel structures*

Let $y(x,t)$ denote the displacement of a continuous elastic system (beam) in space-time coordinates. The modal representation of the above displacement field reads:

$$y(x,t)/D=\sum_{r=1}^{\infty}n_r(t)\varphi_r(x) \tag{1}$$

where $n_r(t)$ denotes the time varying modal coordinate or modal amplitude and $\varphi_r(x)$ the modeshape or modal function, while D is a characteristic length used for normalizing the displacement field.

The equation of motion governing the elastic problem in modal co-ordinates yields:

$$\ddot{n}_i(t)+2\xi_i\omega_i\dot{n}_i(t)+\omega_i^2 n_i(t)=\frac{F_i(t)}{DM_i} \tag{2}$$

where ω_i and ξ_i are the modal frequency and modal damping ratio respectively of the ith mode, while M_i and $F_i(t)$ are the corresponding modal mass and modal force, defined as:

$$M_i=\int_0^l m(x)\varphi_i^2(x)dx=m_{ei}\int_0^l \varphi_i^2(x)dx \tag{3}$$

$$F_i(t)=\int_0^l f(x,t)\varphi_i(x)dx \tag{4}$$

where $m(x)$ is the linear mass per unit length, m_{ei} is the equivalent mass per unit length and $f(x,t)$ is the force per unit length.

In the simple case of a cantilever cylindrical beam of constant external diameter D, the VIV load can be defined as a sinusoidal crossflow lift force per unit length given by:

$$f(x,t)=\frac{1}{2}\rho V^2 D\Delta C_l \cos(\omega t) \tag{5}$$

where ρ is the air density, V is the wind speed, ΔC_l is the sinusoidal lift coefficient amplitude and ω the angular velocity. The elastic system gets into resonance as ω approaches one of its modal frequencies ω_i. When resonant lock-on conditions occur, vortices are shed coherently over a large area of the cylinder, the correlation length L_i referring to the ith mode. The RHS of the equation of motion (2) is then written as,

$$\frac{F_i(t)}{DM_i}=\frac{1}{2}\rho V^2 \Delta C_l \cos(\omega t)\frac{1}{m_{ei}}4\pi K_w K = 2\xi_i \omega^2 \frac{1}{Sc}\frac{1}{St^2} K_w K\ \Delta C_l \cos(\omega t) \tag{6}$$

where K_w is the effective correlation length factor and K is the mode shape factor, given by [7]:

$$K_w = \int_{L_i} |\varphi_i(x)|dx \Big/ \int_0^l |\varphi_i(x)|dx \tag{7}$$

$$K = \frac{1}{4\pi}\int_0^l |\varphi_i(x)|dx \Big/ \int_0^l \varphi_i^2(x)dx \tag{8}$$

In the second expression in (6) the Strouhal (St) and the Scruton (Sc) numbers appear, defined as:

$$St = \frac{D\omega}{2\pi V} \tag{9}$$

$$Sc = \frac{4\pi\xi_i m_{ei}}{\rho D^2} \tag{10}$$

Following simple dynamic systems' theory, the maximum normalized displacement n_i at resonance is calculated as,

$$max(y_i / D) = \max(n_i) = \frac{1}{Sc}\frac{1}{St^2}K_w K \Delta C_l \tag{11}$$

The above expression is the formula provided by the EUROCODE for "Approach 1". It is noted that the maximum displacement increases with K_w suggesting that VIV is more pronounced when the coherent shedding takes place at the antinode of the considered modeshape, where the numerator of (7) attains its maximum. Moreover, the maximum displacement increases as the Scruton number gets lower, which, for a given aeroelastic system, happens when the total modal damping (structural, aerodynamic, soil, etc.), expressed through ξ_i, gets lower. In the opposite, an increased modal mass has stabilizing effect on VIV.

"VIV Approach 1" is applicable for any mode of generalized space frames consisting of beam elements due to its generality [7]. Its main drawback is that it lets all the VIV lock-on physics to be expressed through the semi empirical framework for L_i and ΔC_l, failing to comply with the strongly non-linear lock-on features that relevant experiments demonstrate.

The semi-empirical formulas suggested by EUROCODE for the calculation of L_i and ΔC_l are provided next for completeness. The non-dimensional correlation length, as a function of the maximum normalized displacement n_i, is given by,

$$\frac{L_i}{D} = min\left(max\left(4.8+12n_i,6\right),12\right) \tag{12}$$

whereas the oscillation lift amplitude ΔC_l is given by,

$$\Delta C_l = \min\left(max\left(3-2.4\frac{V_{crit,i}}{V_{m,Li}},0\right),1\right)C_{lat,0} \tag{13}$$

where $C_{lat,0}$ is the basic value of ΔC_l as given in Table E.2 of the EUROCODE, which for cylindrical bodies is assumed equal to 0.2 for the Reynolds numbers within [5 10^5 , 5 10^6], $V_{crit,i}$ is the critical wind velocity, given as a function of the Strouhal number and $V_{m,Li}$ is the mean velocity in the center of the correlation length L_i.

2.2. Additional negative aerodynamic damping term in VIV forcing term

In the following, we extend "Approach 1" theory with some successful features of "Approach 2". EUROCODE provides a second, more elaborate "Approach 2" for VIV calculations based on [12], the use of which is however restricted to the first mode of cantilever structures with a regular distribution of cross wind dimensions along the main axis of the structure. This approach allows for the consideration of different turbulence intensities, which may differ due to meteorological conditions. The essence of the theory behind "Approach 2" is the addition of an extra term to the aerodynamic force, which introduces negative aerodynamic damping when the structure starts vibrating because of vortex shedding. The aerodynamic force in equation (5) is then written as:

$$f(x,t)=\frac{1}{2}\rho V^2 D\ \Delta C_l \cos(\omega t)+2\ \omega_i\ \rho\ D^2 K_a \dot{y}(x,t)\left(1-G\dot{n}_i(t)^2\right) \tag{14}$$

where K_α is the aerodynamic damping coefficient, function of the Reynolds number, the geometry (cross section shape), the turbulence intensity and the stiffness of the structure, and G is a limiter of the negative damping term to avoid the system becoming unstable, so that $n_i(t)$ would never exceed a given value $(y/D)_{lim}$. Limiter G is given by:

$$G=\frac{4}{3}\frac{1}{\omega_i^2 (y/D)_{lim}^2} \tag{15}$$

By introducing (14) in (2) and moving the additional (negative) damping term to LHS, the total damping term of the system is given by,

$$2\xi_i\omega_i\left[1-4\pi\frac{K_a}{Sc}\left(1-G\dot{n}_i(t)^2\right)\right]\dot{n}_i(t) \tag{16}$$

where all the remaining terms of the modal equation of motion (2) and the forcing term (6) remain unchanged. An approximate algebraic expression for $max(y_i/D)$ can be obtained for the derived non-linear dynamic equation using symbolic mathematics language.

In case no limiter is applied (i.e. $G=0$), the system becomes unstable when its total damping gets negative, which happens for $Sc\leq 4\pi K_\alpha$. However, this would not respect the physics of a dynamic system in resonance, whose extreme response is limited and not perpetually diverging.

In the above, one has still to calibrate the driving parameters L_i and ΔC_l but also K_α and $(y/D)_{lim}$. Nevertheless, this time L_i and ΔC_l do not have to include lock-on effects as the lock-on physics are now expressed in a phenomenological way through the limited damping term. Therefore, the L_i and ΔC_l values can be directly taken from aerodynamic (i.e. omitting the aero-elastic coupling) experiments or calculations. Also, the exact value of $(y/D)_{lim}$ is not that essential since one must prevent the structure entering in the negative damping region, which is signified by Sc and K_α, while the system's response around the critical Scruton number is insensitive to $(y/D)_{lim}$.

Higher turbulence levels reduce the critical Scruton $Sc_{crit}=4\pi K\alpha$ under which the vibrations get stronger. In this respect, K_α is expressed as $K_\alpha= K_{\alpha,max}\ h(TI)$, where TI expresses the fraction of the turbulence spectrum which lies above a characteristic for the structure length scale and $h(TI)$ is a declining function of TI. A simplified linear expression for h is $h(TI)=1-3TI$. For the range of application of "Approach 2", for circular cylinders and Reynolds numbers $Re\geq 10^6$, $K_{\alpha,max}=1$.

2.3. Extension to WT arrangements

Applying both "Approaches" to multiple turbine configurations with different number of blades and comparing vortex shedding results against an "equivalent" tower – lumped RNA analogue, it appeared that for unification reasons it was better to work with a modified Scruton number Sc^*, instead of the original Sc. This is because as additional bodies (blades) are considered in the system, m_{ei} but also K_w

values change drastically because their defining relations integrate the mode shapes along the added bodies as well. As a result, the bladed system and the equivalent RNA mass-lamped system would significantly differ in critical Scruton numbers (because of the m_{ei}) in the case of tower VIV, which is not reasonable. The modified Scruton is defined as,

$$Sc^* = Sc/(4\pi K_w K) \tag{17}$$

whereas (11) takes the form,

$$max(y_i/D) = \max(n_i) = \frac{\Delta C_l}{4\pi}\frac{1}{Sc^*}\frac{1}{St^2} \tag{18}$$

The critical Sc^* value becomes $Sc^*_{crit} = \beta K_a = \beta K_{a,max}(1-3TI)$, with $\beta = 1/(K_w K)$. For circular cylinders vibrating at their 1st mode $\beta \cong 13-15$.

Further, the above formulas have been obtained using the same characteristic length D to dimensionalize maximum displacement $max(y_i/D)$ and to define the Strouhal (St) and Scruton (Sc) numbers. In practice for the engineering semi-empirical framework established in the present work we needed to include blade/tower tapering and the case that vortex shedding springs from the blades but excites the tower modeshape. In this case, two distinct characteristic lengths should be considered, one for defining Strouhal denoted by C and another to dimensionalize the displacement denoted by D as previously. Then equation (18) for the maximum normalized displacement takes its final form,

$$max(y_i/D) = \max(n_i) = \frac{\Delta C_l}{4\pi}\frac{1}{Sc^*}\frac{1}{St^2}\frac{C^3}{D^3} \tag{19}$$

2.4. *Assumptions and modelling details*

Our basic assumption is that the well-known Crossflow or Independence Principle (IP) applies to wind turbine blades too. IP suggests that the yawed geometry, receiving the flow with an inclination angle, behaves like the normal-incidence case using the component of the free-stream velocity normal to the body axis. This is important in our case for coping with 2-bladed and 3-bladed configurations, where the 'Λ or V' blades see the flow with a ±60° inclination in respect to the blade stacking axis.

In Kaja et al. [13], the wake of tapered cylinders for certain inflow conditions accounted for a combination of two distinct shedding patterns (i.e., 2S and 2P). The lock-in regime of the reduced velocity for a tapered cylinder was found to be wider than that of a uniform cylinder. This is mainly because the variation of the local reduced velocity along the cylinder span increases the chance of lock-in, compared with the single reduced velocity in the uniform cylinder case. In addition, although the vortex shedding pattern may vary along the cylinder span, the frequencies of the vortex shedding and the sectional lift coefficient synchronize with the vibration frequency of the cylinder in the lock-in regime instead of varying along the cylinder span. Outside the lock-in regime, the flow pattern is like that of a single cylinder due to very small vibration amplitude. In our semi-empirical framework we apply the IP principle on tapered blade/tower geometries with the reduced velocity calculated using the mean characteristic length (chord or diameter) of the vortex shedding region.

The Strouhal number of the shed vortices depends on the cross section of the body. EUROCODE suggests St=0.18 for cylindrical cross sections but provides no value for airfoils. Horcas et al. [6] suggested a projected chord-based St number of 0.15-0.16 for inboard triggered vortex shedding. In an earlier work of ours [2] we have estimated St values for 2D airfoils in the range of 0.11-0.13, also reported in [3]. For the quantification of maximum displacements, we shall use St=0.15 for inboard shedding and St=0.12 for outboard shedding.

Lucor and Karniadakis [14] found that, for low Reynolds flows, the energy in the Strouhal peak disperses and decreases significantly with increasing inclination θ, whereas it is virtually submerged in the general wake turbulence spectrum for θ ≥ 60°. For θ = 60°, the rms values of the lift coefficient peak at the reduced velocity correspond to the largest cylinder response. Therefore, it seems that there is a loss of spanwise coherence and a drop in the correlation of the lift forces when the cylinder's response amplitude is maximum. We account for this effect in our semi-empirical framework by reducing the coherence length of the inclined shedding area by 25% when $\theta = 60°$, with the coherence length tending to zero as θ tends to 90° (currently applied only in case the rotor is considered locked).

In their paper Horcas and al. notice that at low inclination angles (0° to 10°) an unsteady turbulent wake dominates the flow, which is characterized by the shedding of disorganized vortical structures. When the inclination angle increases, stationary vortices are observed for a certain extent of the outboard part of the blade pointing towards the wind. The leeward vortices of this stationary wake are convected away from the blade following the flow direction forming a set of straight vortical tubes. At 70° inclination, the stationary wake is seen all along the blade span, remaining close to the body surface, while the shedding is only taking place in the vicinity of the blade root. In addition, the results of the flexible simulations of the baseline blade geometry showed significant edgewise vibrations for inclination angles between 47.5° and 60°. This lock-in condition between the fluid loading and the blade displacements was always accompanied by a reconfiguration of the stationary wake region into a parallel vortex shedding. This shedding was also observed at the tip vicinity, and it followed a 2S pattern.

Summing up for this tip effect it appears that a stiff blade in 'Λ or V' configuration with its tip pointing to the wind does not shed vortices from its tip region and one can make a similar hypothesis for the root region of its companion blade. Nevertheless, there is evidence that when VIV is triggered and the structure starts vibrating, this behaviour may change. For our purposes, and trying to avoid extreme conservatism, we shall assume that the two 'Λ or V' blades cannot shed vortices simultaneously both from their tip or their root region. The above assumption is further supported in case the rotor is free, by the fact that the two blades encounter different 2D projected velocities due to asymmetry in azimuth angle when the wind is coming from the side direction and the blades are pitched to feather position.

3. Numerical results and discussion

Numerical results for single- and two-bladed configurations of the NREL 5MW RWT [11] during assembly are presented using the established VIV simulation framework, assessing turbine loads under the most unfavorable VIV scenarios examined. The rotor is in idling mode (i.e. the brake is disengaged) and the blades pitched to feather position at 90°. Structural damping values of 0.19% and 0.25% are used for the tower and the blades respectively, significantly lower that the baseline values reported in [11], adapting to state-of-the art damping levels.

Table 1 provides the first and second stand still natural frequencies of the coupled NREL 5MW RWT configuration, equipped with one, two or three blades at 90° pitch, with free rotor (free-free drivetrain) and the gravity effect included. In the single- and the two-bladed configuration the rotor is analyzed at its static equilibrium state with the blades pointing downwards (one blade is vertical at 180° azimuth and two blades in 'Λ' position at 120° azimuth), while in the three-bladed configuration the first blade points upwards (0° azimuth). In the table, relative percentage differences are shown with respect to the three-bladed (full) configuration, which is included as a means of comparison although not analyzed.

Since the number of (first, second and higher) rotor modes per bending direction (i.e. flapwise and edgewise) should be equal to the number of the installed blade(s), one mode (symmetric) appears in the single-bladed case and two modes (symmetric and one asymmetric) in the two-bladed configuration per direction. Minor frequency differences are depicted that do not exceed 5% and 10% in the 2-bladed and 1-bladed configuration respectively. Expected tower frequency increase is depicted (~8% and 5%) in case one or two blades are equipped due to the reduced RNA mass. Higher frequency differences are seen in both first symmetric modes of the 1-bladed (about -10%) and 2-bladed (-4%) configurations, whereas the frequency difference of the remaining modes is marginal.

Modal analysis results define the modeshapes to be checked for VIV. The low damped first tower fore-aft and rotor edgewise mode(s) are expected to be prone to VIV when the blades are pitched to feather. Second modes of the analyzed configurations are too high in frequency to be excited at moderate to high wind speeds (2nd tower frequencies ~2.8Hz), so they will not be included in VIV analysis. In addition, modal analysis indicates the most effective for VIV shedding areas per modeshape. As mentioned in section 2 and demonstrated in Figure 1 for the single-bladed case (similar is the two-bladed case), shedding should be as close as possible to the antinode of the (excited) modeshape. In this respect, shedding from the tower top and the blade tip areas is expected to be worst case for the tower fore-aft and rotor edgewise mode(s) respectively. For tower mode excitation, two possible scenarios are considered, a) shedding from tower top and b) shedding from blade root.

Table 1: Stand still natural frequencies [Hz] of the coupled NREL 5MW RWT equipped with one, two and three blades with 90° pitch, free rotor and gravity included. Relative frequency differences with respect to the full (3-bladed) configuration are also provided.

	Modeshape	3-bladed	2-bladed		1-bladed	
1	1st tower fore-aft	0.32	0.33	5%	0.34	8%
2	1st tower side-side	0.32	0.33	3%	0.34	7%
3	1st rotor flapwise asymmetric 1	0.69	0.69	0%	-	-
4	1st rotor flapwise asymmetric 2	0.69	-	-	-	-
5	1st rotor edgewise asymmetric 1	0.88	0.88	0%	-	-
6	1st rotor edgewise asymmetric 2	1.02	-	-	-	-
7	1st rotor flapwise symmetric	1.04	1.00	-4%	0.94	-10%
8	1st rotor edgewise symmetric	1.16	1.11	-4%	1.06	-8%
9	2nd tower fore-aft	2.72	2.79	2%	2.78	2%
10	2nd tower side-side	2.88	2.88	0%	2.87	0%
11	2nd rotor flapwise asymmetric 1	1.97	1.97	0%	-	-
12	2nd rotor flapwise asymmetric 2	1.97	-	-	-	-
13	2nd rotor flapwise symmetric	2.04	2.03	-1%	2.01	-2%
14	2nd rotor edgewise asymmetric 1	2.77	2.79	1%	-	-
15	2nd rotor edgewise asymmetric 2	3.92	-	-	-	-
16	2nd rotor edgewise symmetric	4.08	3.99	-2%	3.93	-4%

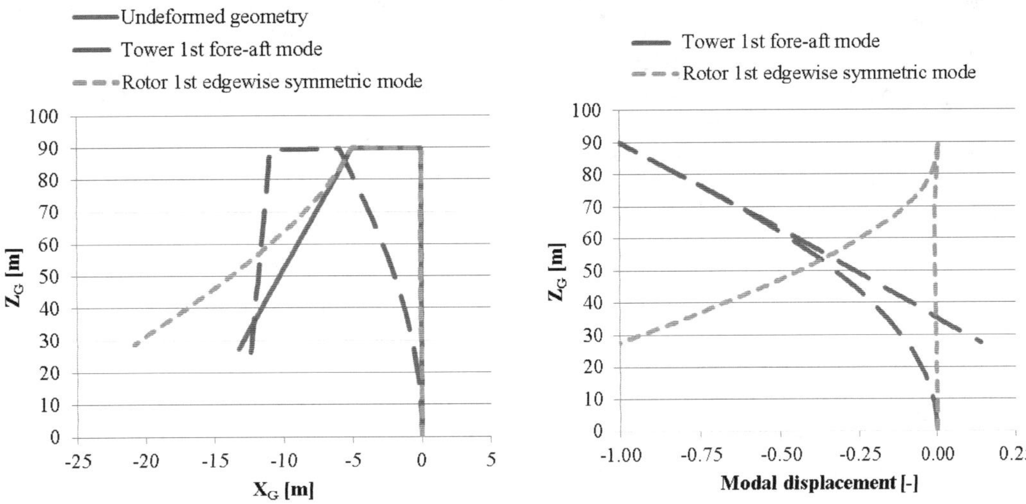

Figure 1: Low damped 1st tower fore-aft and 1st rotor (symmetric) edgewise modeshapes considered in VIV analysis of the NREL 5MW RWT single-bladed configuration; side view of the global 3D representation (left) and modal shape versus height indicating the antinodes at which modal displacement is maximum (right).

Before applying EUROCODE VIV semi-empirical framework, eigenvalue stability calculations are performed with the eigenvalue stability module of hGAST for various wind speeds for wind speeds in the range of 5m/s to 30m/s to obtain the modal characteristics of the low damped modeshapes of interest, namely the natural frequency, modeshape, steady state azimuth angle of the free rotor and total (structural plus aerodynamic) modal damping. Steady-state aerodynamics is assumed because for the yaw angles around ±90°, which are considered in VIV analysis we lack a reliable dynamic stall model to be used instead. Further, the effect of the vortex shedding is not taken into consideration. Steady wind inflow (i.e. no turbulence) with wind inclination of 8° and shear exponent of 0.11 has been selected.

Figure 2 shows the steady state azimuth angle of the two configurations, which gradually deviates from the equilibrium angle (i.e. 120° and 180° for the two- and single-bladed configuration respectively) as the wind speed increases. In both cases, it is seen that the rotor cannot rotate for any of the wind speeds considered with steady inflow (i.e. without turbulence).

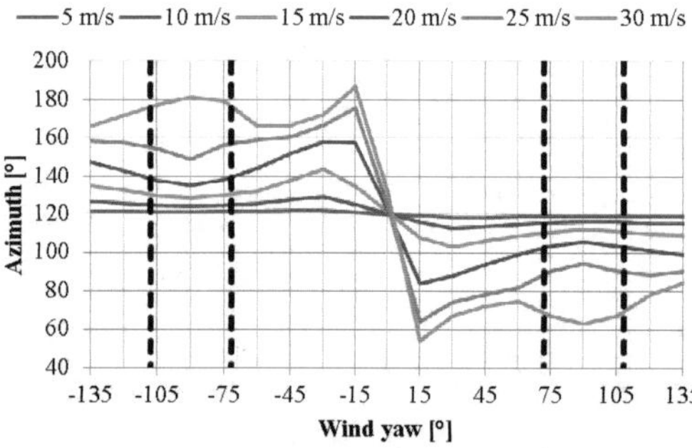

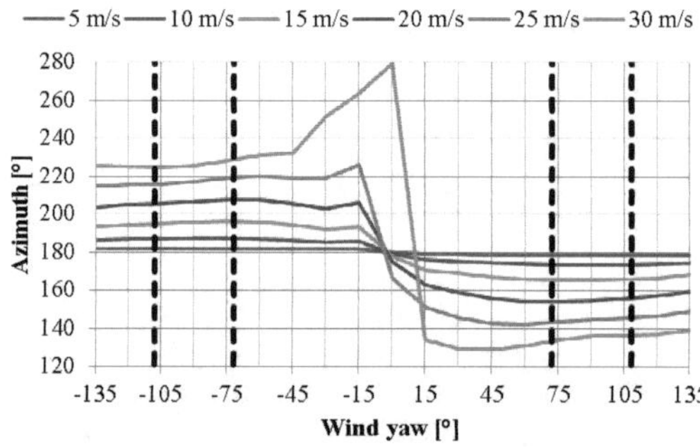

Figure 2: Rotor azimuth angle versus yaw of the NREL5 MW RWT two-bladed (left) and single-bladed (right) configurations for various wind speeds with free rotor, steady wind, wind inclination 8° and shear exponent 0.11, resulting from the eigenvalue stability analysis.

Figure 3 provides the total modal damping values for yaw angles around ±90°, which will be introduced in tabulated form in the VIV analysis semi-empirical framework. As indicated with the dashed black lines, we restrict the VIV analysis at flow conditions that the blades face the flow at angles of attach within the ranges [-105°, -75°] and [75°, 105°]. Similar damping values are estimated for the same modes of the singe- and the two-bladed configuration. Tower total modal damping is about 0.19-0.20%, almost insensitive to the wind speed variation, which means that the aerodynamic damping contribution is negligible. The aerodynamic damping contribution on the rotor edgewise asymmetric mode in the two-bladed case is small, as the total modal damping spans from 0.25% to 0.60% for the wind speeds considered. The rotor symmetric edgewise mode is higher damped as demonstrated through the increase of the damping values for increasing wind speeds. It should be noted that the variation of the azimuth angle of the free rotor introduces an inherent geometric non-linearity, so that damping levels are not linearly increasing with wind speed.

Results of the VIV analysis are presented in the next three tables. The 'VIV approach 1' is employed (i.e. K_α=0). It is reminded that shedding from one blade of the two-bladed configuration is only considered (the worst case is shown), whereas $\Delta C_l = 0.2$ and air density of 1.25kg/m^3 is assumed. Table 2 shows the input parameters of the worst VIV cases defined. *Case* is the type of shedding, first the body that sheds vortices and then the modeshape that is excited. For example B-T means shedding from the blade at the tower frequency, *Mode* is the excited mode, *Shedding* is the shedding body area of the assumed "worst case scenario" (where K_w is maximized), *St* is the imposed Strouhal number, *D* is the reference length (diameter or chord) used for the non-dimensionalization of the maximum oscillation deflection amplitude and *C* is the mean characteristic length of the shedding area.

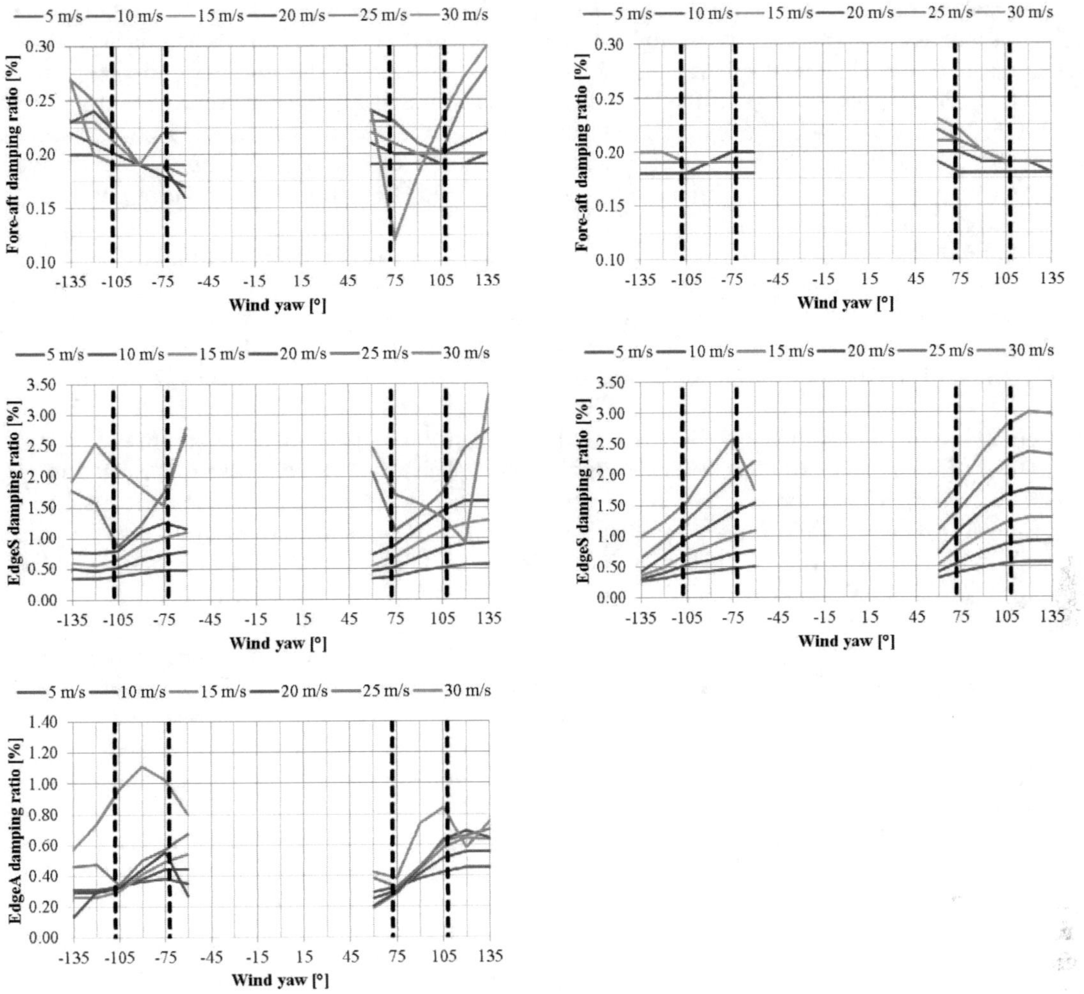

Figure 3: Total (structural & aerodynamic) modal damping ratio of the first fore-aft tower (top), rotor edgewise symmetric (middle) and rotor edgewise asymmetric (bottom) modes versus yaw of the NREL 5MW RWT two-bladed (left) and single-bladed (right) configurations for various wind speeds, resulting from the eigenvalue stability analysis with steady-state aerodynamics.

Table 2: Definition of the 'worst' cases and input parameters considered in VIV analysis.

ID	Case	Mode	Shedding	St	D	C
[-]	[-]	[-]	[-]	[-]	[m]	[m]
1	T-T	fore-aft	Tower Top	0.18	3.87	3.87
2	B-T	fore-aft	Blade 1 Root	0.15	3.87	4.20
3	B-BA	edgeA	Blade 2 Tip	0.12	1.40	2.40
4	B-BS	edgeS	Blade 1 Tip	0.12	1.40	2.40

Table 3 and Table 4 provide the calculated modal quantities defined in the EUROCODE framework (see section 2 for their definition) and the output of the VIV analysis for the single- and the two-bladed configuration respectively. f and ξ denote the excited modal frequency and total damping ratio at the "worst case scenario" (minimum aerodynamic damping value within the considered yaw misalignment regime) derived through interpolation from the stability analysis results. *NL* is the non-dimensional metric denoting the extent of the shedding area in characteristic length L multiples which derive through equation (12), m_e is the equivalent mass per unit length, K is the modeshape factor, K_w is the effective correlation length factor, while Sc and Sc^* denote the Scruton and the modified Scruton number respectively. V_{crit} is the Critical 2D cross sectional wind speed corresponding to the St number considered, V_{inf} is the actual wind velocity which projected on the shedding body (based on IP) yields V_{crit}, YAW_{inf} is the yaw misalignment of the "worst case scenario", *Azim* is the azimuthal position of the free rotor and Y_{max} is the maximum combined oscillation deflection amplitude of the considered antinode.

It can be seen that apart from K all other modal quantities are significantly different between the two configurations. Specifically, K_W, m_e and Sc are almost double in the single-bladed configuration as compared to the two-bladed one. On the contrary, Sc^* values are very similar in case the oscillation amplitudes (Y_{max}) are close, justifying our choice to define it.

T-T case #1 represents tower excitation due to tower-top vortex shedding. The "worst case scenario" occurs for the single-bladed turbine at V_{inf}=7.3 m/s and YAW_{inf}=-105° and for the two-bladed configuration at V_{inf}=7.1 m/s and YAW_{inf}=-75°, whereas the maximum oscillation amplitude is 0.08 m and 0.07 m respectively. B-T case #2 represents tower excitation due to blade-root shedding. The "worst case scenario" for the single-bladed rotor occurs at V_{inf}=9.6 m/s (due to the difference in C_{ref} compared to case #1), while for the two-bladed setup at V_{inf}=14.7 m/s due to the projection of the inflow wind speed (based on IP), whereas YAW_{inf} is the same as case #1. The maximum oscillation amplitude of the single- and two-bladed case is 0.16 m and 0.18 m respectively. B-BA case #3 represents blade edgewise asymmetric mode excitation due to blade 2-tip shedding. The "worst case scenario" occurs at V_{inf}=17.6 m/s and yaw misalignment 75°, whereas the maximum combined oscillation amplitude is 2.61 m, a rather high value. Lastly, B-BS case #4 represents blade edgewise symmetric mode excitation due to blade-tip shedding. The "worst case scenario" occurs at yaw misalignment -105° and V_{inf}=28.5 m/s and 24.8 m/s for the single- and the two-bladed rotor respectively, whereas the maximum combined oscillation amplitude is 1.06 m and 1.34 m.

Table 3: VIV analysis results of the single-bladed NREL 5MW RWT configuration.

ID [-]	NL [-]	f [Hz]	ξ [%]	m_e [kg/m]	K [-]	Kw [-]	Sc [-]	Sc* [-]	V_{crit} [m/s]	V_{inf} [m/s]	YAW_{inf} [°]	Azim [°]	Y_{max} [m]
1	6.0	0.339	0.18	8087	0.13	0.27	9.8	23.1	7.3	7.3	-105	-	0.08
2	6.0	0.339	0.18	8087	0.13	0.29	9.8	21.0	9.5	9.6	-105	186	0.16
3	-	-	-	-	-	-	-	-	-	-	-	-	-
4	12.0	1.062	1.46	105	0.14	0.60	7.9	7.4	21.2	28.5	-105	222	1.06

Table 4: VIV analysis results of the two-bladed NREL 5MW RWT configuration.

ID [-]	NL [-]	f [Hz]	ξ [%]	m_e [kg/m]	K [-]	Kw [-]	Sc [-]	Sc* [-]	V_{crit} [m/s]	V_{inf} [m/s]	YAW_{inf} [°]	Azim [°]	Y_{max} [m]
1	6.0	0.332	0.18	3570	0.11	0.13	4.3	24.2	7.1	7.1	-75	-	0.07
2	6.0	0.332	0.19	3570	0.11	0.17	4.5	19.7	9.3	14.7	-75	129	0.18
3	12.0	0.879	0.30	121	0.13	0.39	1.9	2.9	17.6	22.2	75	218	2.61
4	12.0	1.112	0.87	97	0.15	0.39	4.3	5.7	22.2	24.8	-105	154	1.34

Figure 4 shows the spanwise distributions of the oscillating bending and deformation amplitudes of the worst VIV cases examined. Only the excited modeshapes are shown per component, thus tower signals correspond to tower mode excitation (VIV cases 1-2) and blade signals to rotor mode(s) excitation (VIV cases 3-4). Blade amplitudes due to tower excitation and tower amplitudes due to rotor mode(s) excitation are not shown, as for the analyzed configurations the couplings are insignificant and thus the amplitudes rather small.

It should be mentioned that it is out of scope of the present work to study whether the estimated load amplitudes are affordable or not and to examine the exact fatigue damage caused. Nevertheless, both tower cases (T-T and B-T) examined seem to be affordable and of low risk. The B-BS case could be considered of intermediate risk and alarming for low cycle fatigue damage as the oscillation amplitude of the blade moments is quite high for both configurations analyzed. Finally, the B-BA case, defined exclusively for the two-bladed rotor, is considered of high risk and very critical not only for fatigue damage but also for survival, as the oscillation moment amplitude approaches or even exceeds the ultimate design value. Consequently, the two-bladed configuration cannot be left unattended as it suffers from severe VIV.

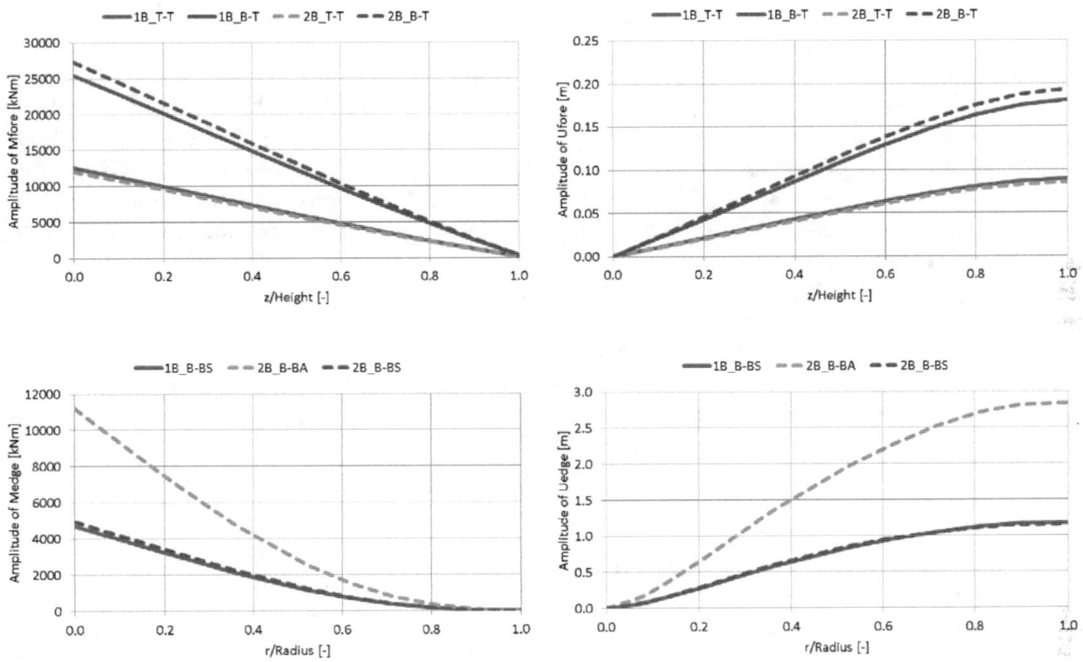

Figure 4: Spanwise tower (top) and blade (bottom) distribution of the oscillating moment (left) and deformation (right) amplitudes of the worst VIV cases examined of the single- and two-bladed NREL 5MW RWT configurations.

4. Conclusions

In the present work, an engineering semi-empirical framework was proposed to assess VIV aero-elastic instabilities of wind turbine configurations. It uses an extended implementation of EUROCODE "Approach 1" VIV framework for wind turbine configurations, which is incorporated in the state-of-the-art aero-elastic tool hGAST. The aero-elastic tool provides the modal input data needed to evaluate the semi-analytical framework. Numerical results for single- and two-bladed configurations of the NREL 5MW RWT during assembly were presented for the worst case VIV scenarios examined. The tower VIV cases examined were found to be affordable and of low risk, whereas the blade VIV cases are considered of intermediate (edgewise symmetric mode) and of high risk (edgewise asymmetric mode), as the oscillation amplitude of the blade edgewise moment is high.

It should be noted that EUROCODE VIV analysis method is trustful within the assumptions of the semi-empirical aerodynamic framework applied. In practice it is known that the method can be quite inaccurate providing less-conservative results when instabilities do exist and more conservative when they are absent. To increase confidence in the results, the aerodynamic framework needs further adaptation and calibration, which is only possible through dedicated high fidelity aeroelastic analysis or experimental measurements.

The present engineering numerical framework can be used to efficiently scan a wide list of critical for VIV cases (e.g. all low damped modeshapes along with various critical shedding areas per mode), defined a-priori, and to provide valuable information about the critical inflow conditions and the corresponding oscillation load and deformation amplitudes. Various configurations prone to VIV might be considered, such as partially installed rotors during assembly and maintenance or parked turbines, as well as various pitch settings with locked or free drive-train. Even more important, although the procedure is subjected to the uncertainty due to the polars and the semi-empirical VIV framework, it can still be used in the design process to assess critical for VIV design parameters and to compare designs.

References

[1] Fontechaa R, Hennekea B, Kempera F and Feldmannb M 2017 Aerodynamic properties of wind turbine towers based on wind tunnel experiments *Procedia Engineering* **199** 3121–3126 DOI:10.1016/j.proeng.2017.09.557

[2] Zou F, Riziotis V A, Voutsinas S G and Wang J 2014 Analysis of vortex-induced and stall induced vibrations at standstill conditions using a free wake aerodynamic code *Wind Energy* **18** 2145-2169 DOI:10.1002/we.1811

[3] Pellegrino A and Meskell C 2013 Vortex shedding from a wind turbine blade section at high angles of attack *J. Wind Eng. Ind. Aerodyn.* **121** 131–137 DOI:10.1016/j.jweia.2013.08.002

[4] Skrzypiński W, Gaunaa M, Zahle F and Heinz J 2014 Vortex-induced vibrations of a DU96-W-180 airfoil at 90 angle of attack *Wind Energy* **17** 1495–1514 DOI:10.1002/we.1647

[5] Heinz J C, Sørensen N N, Zahle F, and Skrzypiński W 2016 Vortex-induced vibrations on a modern wind turbine blade *Wind Energy* **19** 2041–2051 DOI:10.1002/we.1967

[6] Horcas S G, Barlas T, Zahle F and Sørensen N N 2020 Vortex induced vibrations of wind turbine blades: Influence of the tip geometry *Phys. Fluids* **32** 065104 DOI:10.1063/5.0004005

[7] Eurocode 1: Actions on structures - Part 1-4: General actions - Wind actions EN 1991-1-4:2005

[8] Riziotis V A and Voutsinas S G 1997 GAST: A general aerodynamic and structural prediction tool for wind turbines *Proc. of the EWEC Conference* (Dublin Castle Ireland) pp 448–452

[9] Manolas D I, Riziotis V A, and Voutsinas S G 2015 Assessing the importance of geometric non-linear effects in the prediction of wind turbine blade loads *J. Comput. Nonlinear Dynam.* **10** 041008 DOI:10.1115/1.4027684

[10] Manolas D I, Riziotis V A, Papadakis G P and Voutsinas S G 2020 Hydro-servo-aero-elastic analysis of floating offshore wind turbines *Fluids* **5** 200 DOI:10.3390/fluids5040200

[11] Jonkman J M, Butterfield S, Musial W, and Scott G 2009 Definition of a 5-MW reference wind turbine for offshore system development National Renewable Energy Laboratory Golden, CO.

[12] Vickery B J and Basu R I 1983 Across-wind vibrations of structures of circular cross-section. Part I: development of a mathematical model for two-dimensional conditions *J. Wind Eng. Ind. Aerodyn.* **12** 49–73 DOI:10.1016/0167-6105(83)90080-6

[13] Kaja K, Zhao M, Xiang Y and Cheng L 2016 Three-dimensional numerical simulations of vortex-induced vibrations of tapered circular cylinders *Appl. Ocean Res.* **60** 1–11 DOI:10.1016/j.apor.2016.08.004

[14] Lucor D and Karniadakis G E 2003 Effects of oblique inflow in vortex-induced vibrations *Flow, Turbul. Combust.* **71** 375–389 DOI:10.1023/B:APPL.0000014929.90891.4d

Influence of blade deflections on wind turbine noise directivity

A P C Bresciani[1,2,3], U Boatto[4,5], S Le Bras[1], P Bonnet[4] and L D de Santana[2]

[1]Siemens Industry Software N.V., Interleuvenlaan 68, Leuven, 3001, Belgium
[2]University of Twente, PO Box 217, Enschede, 7500 AE, the Netherlands.
[3]Centre Scientifique et Technique du Bâtiment (CSTB), 24, rue Joseph Fourier, 38400, Saint Martin d'Hères, France
[4]Siemens Industry Software N.V., Rue des Chasseurs Ardennais 8, Angleur, 4031, Belgium.
[5]Delft University of Technology, Department of Aerodynamic Wind Energy and Propulsion, Kluyverweg 1, 2629 HS, Delft, the Netherlands.

E-mail: a.p.c.bresciani@utwente.nl

Abstract. This paper establishes the effect of blade deflections on wind turbine noise directivity. Fast turn-around methods are used in a framework of integrated aeroelastic and aeroacoustic simulations: the blade element momentum theory is coupled with a RANS-informed Amiet's model for the aeroacoustic modelling of trailing- and leading-edge noise. This approach is applied to the NREL 5 MW wind turbine and the results of rigid and flexible blades are compared. The overall sound pressure level computed with flexible blades increases up to 13 dBA for listeners close to the rotor plane. This effect is attributed to the flapwise angular deflection of the wind turbine blade. Furthermore, the symmetry of the results with respect to the rotor plane is lost when the flapwise deflection is considered, indicating that the modelling of this rotation is of fundamental importance for the acoustic simulation.

1. Introduction

Large blades installed in modern horizontal axis wind turbines experience significant deflections within the operating life. Recent simulations reproducing the rated power production conditions predicted axial deflections ranging from 8 to 9% of the rotor radius for a 5 MW-class turbine [1,2], around 10% for a 10 MW-class [3], and significantly more than 10% for a 15 MW-class wind turbine [4].

Few studies established the relation between the rotor aeroelasticity and changes on the wind turbine noise emissions. This relation is of particular interest in view of the known extent of deflections for modern large wind turbines. The effect of blade deflections on acoustic results has been studied by Kim et al. [5] with a vortex lattice method, non-linear composite beam theory and semi-empirical models for the noise sources. However, the change in the noise directivity has not been addressed in much detail. Kaviani and Nejat [6] performed Detached Eddy Simulation (DES) coupled with a finite element method to model the blade deformation of a 70 m diameter, 1.5 MW wind turbine. Only a single point spectrum computed with Ffowcs-Williams and Hawkings (FW-H) technique has been provided, giving little insight into the acoustics effect of flexible blades.

In the present work the effect of the blade deflections on the noise directivity is studied considering the NREL 5 MW wind turbine [7]. This 126-meter diameter wind turbine is representative of modern onshore wind turbines. For wind turbines in design operating conditions, trailing-edge noise has been recognised as the most important noise contribution [8]. In this work the leading-edge noise contribution is also considered, highlighting under which conditions it becomes the dominant noise source. Semi-analytical and semi-empirical aeroelastic and aeroacoustic prediction methods are chosen to investigate several operating conditions and to compare the noise radiated by flexible and rigid blades. They are based on the Blade Element Momentum Theory (BEMT) and Amiet's theory [9, 10]. The use of semi-analytical and semi-empirical models has some limitations since these models rely on simplifying assumptions and are calibrated against experimental data which can lead to poor prediction for other flow conditions or blade profiles, e.g., attached blade flow, thin airfoils, inviscid wake flow. However, they lead to fast numerical simulations compared to high-fidelity methods, e.g., direct noise computation or DES coupled with FW-H as in reference [6], which provide accurate results but are often too computationally expensive to be applied to industrial geometries and different operating conditions. In this study a Reynolds Averaged Navier–Stokes (RANS)-informed Amiet's theory investigated by [11, 12] for 2D airfoils and wind turbines is applied to the NREL 5 MW wind turbine. An extension of the methodology is proposed to account for the blade deformation effects.

The numerical method used to compute wind turbine noise accounting for blade deformation is presented in section 2. In section 3, the Blade Element Momentum Theory (BEMT), the RANS approach and the Amiet's model are validated for the NREL 5 MW wind turbine. The aeroacoustic results are reported in section 4. The conclusions of this paper are summarised in section 5.

2. Methodology

A diagram of the methodology adopted in this study is shown in fig. 1. For the acoustic simulation, the 3D CAD of the blade is divided into segments in the spanwise direction using Simcenter 3D software [13]. For each segment, the mid-span airfoil is used in a 2D RANS simulation necessary to compute the boundary layer parameters for the trailing-edge noise Amiet's theory. The blade segments have a constant aspect ratio $\lambda = L_{\text{span}}/c_{\text{mean}}$, where L_{span} is the spanwise extension of the segment and c_{mean} is the chord length of the airfoil in the middle of the segment. Since, in general, the blade cannot be exactly divided into segments with constant λ, the segment closer to the root is enlarged up to the first cross-section of the blade with an airfoil shape. For the NREL turbine hereby considered, the blade is divided in 5 segments: the 4 outer segments have a constant aspect ratio $\lambda = 3$, while the one closer to the root has $\lambda = 4.5$ to include the blade up to $R = 11.75$ m, where the first airfoil-shaped cross-section is located. The part comprised between the rotor centre and $R = 11.75$ m has been neglected for the acoustic simulation because the circular cross-section is not suited for the application of Amiet's theory. Furthermore, the inner part of the blade is known to have a small contribution the noise due to its low velocity [14]. The value $\lambda \geq 3$ is chosen to satisfy the large aspect ratio assumption required by the formulation of Amiet's theory used in the present work. The BEMT is used for the aeroelastic simulations and detailed in section 2.1. The real angle of attack including the pitchwise deflection and the induced angle of attack are linearly interpolated at the radial positions defined for the acoustic simulations and used for the 2D RANS simulations. The pitchwise and flapwise deflections are used to compute the free-field noise directivity with Amiet's theory. For the flapwise deflection, the blade displacement is neglected and only the rotation of the blade section is considered.

The methodology described above can be applied with an arbitrary wind speed profile to simulate the atmospheric boundary layer: the rotor disk must be discretized also in the azimuthal direction to account for the varying wind speeds and a 2D RANS simulation should be performed for each azimuthal position and blade segment. In the present work, the wind speed is supposed to be uniform over the rotor disk to obtain axisymmetric conditions and, so, drastically reduce the number of 2D RANS simulations necessary. As shown by Tian and Cotté [15], the modelling of the wind shear has negligible effects on the trailing-edge noise, which, as shown in section 4, is the main noise source up to the rated conditions. For this reason, the conclusions of the present work are not expected to change if the effect of the wind shear is included.

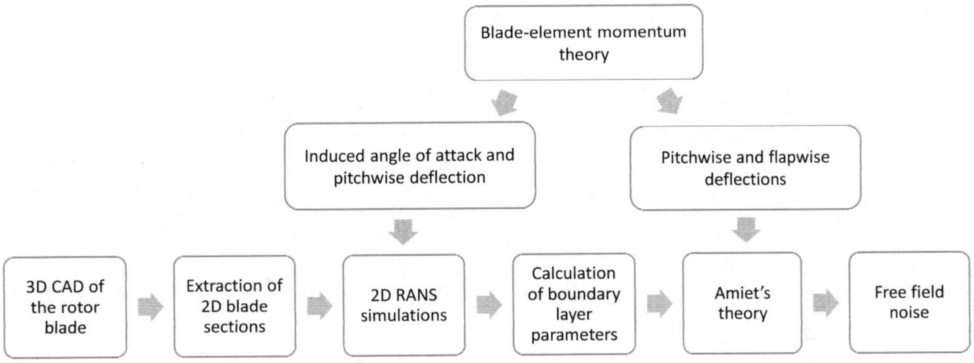

Figure 1: Diagram of the numerical workflow

2.1. *Blade element momentum theory*

In this study, a flexible multi-body simulation is performed with the Simcenter Samcef Mecano [16] software. The solver combines a finite element discretization of the Timoshenko beam theory for large displacements and rotations with the non-linear multi-body framework. This approach is suited to the simulation of vibrating flexible structures within mechanisms subject to complex loads, such as wind turbine blades [17]. The blade aerodynamic forces are computed by a BEMT-based load element that includes all the state-of-the-art engineering corrections employed in similar wind turbine aero-servo-elastic industrial codes [18, 19]. In the present study, the employed BEMT solution consists of the quasi-static Glauert's BEMT theory [20] corrected by the Burton's version [21] of the Prandlt tip and root loss factor [22] and the Buhl's turbulent wake state correction [23].

2.2. *RANS-informed Amiet's theory*

Amiet [9, 10] developed a semi-analytical theory to predict leading- and trailing-edge airfoil noise. The extension of this model to rotating blades is due to Schlinker and Amiet [24] and it is known in literature as strip theory: the blade is divided in strips to account for the different relative velocities and Amiet's theory for leading- and trailing-edge noises is applied to each strip. This model has been successfully applied to predict the noise generated from different types of rotors, such as fans [25–29], helicopters [24] and wind turbines [12,15,30,31]. In Amiet's theory, the airfoil is modelled as an infinitely thin flat plate with chord $c = 2b$, span $L = 2d$ and at zero angle of attack with respect to the freestream velocity U. If the observer is located in the acoustic and geometrical far-field and the aspect ratio L/c of the flat plate is large, the power spectral density for trailing-edge noise S_{pp}^{TE} is computed as [32]:

$$S_{pp}^{TE}(\mathbf{x},\omega) = \left(\frac{\omega x_3 b}{2\pi c_0 \sigma_0^2}\right)^2 4d|\mathcal{I}(x_1,k_{10},k_{20})|^2 l_y(k_{10},k_{20})\Phi_{pp}(k_{10},k_{20}), \tag{1}$$

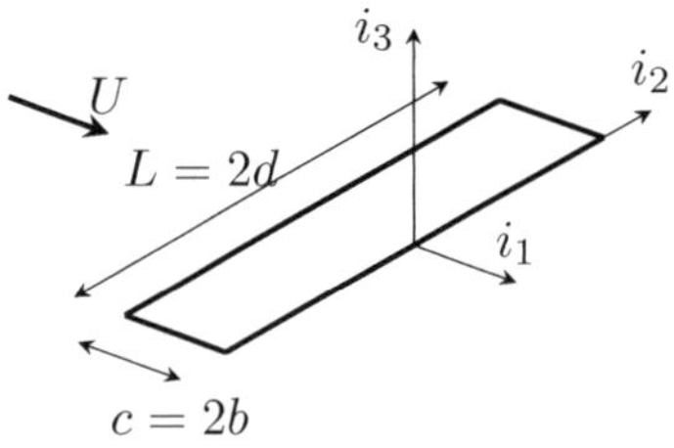

Figure 2: Reference frame for trailing-edge noise.

where $\mathbf{x} = (x_1, x_2, x_3)$ is the position of the observer in the reference frame depicted in fig. 2; ω is the pulsation; c_0 is the speed of sound; $\sigma_0 = \sqrt{x_1^2 + \beta^2(x_2^2 + x_3^2)}$; $\mathcal{I} = \mathcal{I}_1 + \mathcal{I}_2$ is the aeroacoustic transfer function for trailing-edge noise, given in [32, 33]; $k_{10} = \omega/U_c$ with $U_c = 0.7U$; $k_{20} = kx_2/\sigma_0$ with $k = \omega/c_0$; l_y is the spanwise correlation length obtained from Corcos model [34]; Φ_{pp} is the wall pressure spectrum computed with empirical models for computational efficiency.

Empirical [35–38] and analytical models [39] for the wall pressure spectrum have been proposed in literature. However, they need as input boundary layer parameters extracted at the trailing-edge, such as the boundary layer thickness, the wall shear stress, the maximum shear stress and the pressure gradient. All these parameters can be extracted from 2D RANS simulations, allowing a trade-off between low computational cost and accuracy. In the present work, Lee's model [37] has been applied in case of adverse pressure gradient, while Goody's [35] has been used for zero or negative pressure gradient. The extraction location is at 99% of the chord on both sides of the airfoil. To apply Amiet's theory, the blade segments are linearized in flat plates. Their orientation in space is defined by two angles: the pitch angle and the cone angle. The pitch is given by the sum of the twist, the collective pitch and the torsional deformation; the cone is computed using the cone angle of the rotor (set to 0 in this work) and the flapwise deflection angle.

The 2D RANS simulations are performed with CFD software STAR-CCM+ [40]. The turbulence model k-ω SST is used and all the simulations are wall-resolved, i.e. with $y^+ < 1$. The γ-transition model [41] is used to model the laminar-to-turbulent flow transition. All the RANS simulation setup, including the meshing process and the computation of the boundary layer parameters, is fully automated in STAR-CCM+.

Amiet's formulation for leading-edge noise reads

$$S_{pp}^{LE}(\mathbf{x}, \omega) = \left(\frac{\rho\omega c x_3}{2c_0\sigma_0^2}\right)^2 \pi U d |\mathcal{L}(x_1, k_{10}, k_{20})|^2 \Phi_{ww}(k_{10}, k_{20}), \tag{2}$$

where ρ is the air density, $\mathcal{L} = \mathcal{L}_1 + \mathcal{L}_2$ is the aeroacoustic transfer function for leading-edge noise [33] and Φ_{ww} is the inflow turbulence spectrum modelled using the von Karman spectrum in this work. The inputs required by the von Karman spectrum are the wind turbulence intensity and integral length scale. Both quantities depend on the location of the wind turbine, the surrounding environment and the time of the day. Since experimental data for comparison are not available, an integral length scale of 120 m and a turbulence intensity of 10% have been chosen as values representative for typical operating conditions [42, 43].

The pitchwise and flapwise deflections are considered as additional rotation matrices for each blade segment. For the pitchwise deflection, the additional pitch angle resulting from the torsion of the blade is used; for the flapwise deflection, the angle considered is comprised between the deformed blade span axis and the rotor plane. It is worth mentioning that, in the acoustic simulation, the displacement in the off-plane direction has been neglected. The noise contributions from each airfoil are averaged in the azimuthal direction following [44] and, finally, summed.

3. Validation

The NREL 5 MW turbine is modelled as an untilted, unconed, isolated rotor similarly as in [2] and by employing the specifications provided in [7]. The 3D blade CAD geometry is generated by a Siemens Samtech and Universidad Nacional del Litoral in-house code using the aeroelastic simulation parameters and airfoil shapes given in reference [7].

Five conditions representative of steady-state power production operations are considered in this study. The operating conditions are shown in table 1 and the turbine is simulated with fully-rigid and fully-flexible blades, leading to a total of 10 simulations. Rigid blades are obtained by introducing rigid bodies into the blade finite element model to achieve zero deflections. The operational conditions are taken from [7] except for the pitch angle of the above-rated case which was modified to obtain a power value as close as possible to the predicted flexible rated power in table 2. Finally, the air density is set to 1.225 kg/m^3 and the speed of sound to 343.0 m/s.

Table 1: Simulation conditions employed in this study for both rigid and flexible blades.

Condition	Wind speed [m/s]	Rotor speed [rpm]	Pitch angle [°]
Below-rated	5.0	7.51	0.0
Below-rated	8.0	9.30	0.0
Below-rated	10.0	11.25	0.0
Rated	11.4	12.10	0.0
Above-rated	14.0	12.10	8.13

3.1. BEMT validation

The aeroelastic performance of the NREL 5 MW isolated rotor is first compared with the reference 3D unsteady RANS simulations performed by Dose et al. [2] for both rigid and flexible blades cases. First, the rotor thrust and power are compared in table 2 at rated conditions. BEMT can predict very well both quantities with an error below 1% in both simulation cases. For the flexible case, the aeroelastic code predicts 5.54 m axial blade tip deflection towards downwind and 0.62 m blade tip tangential deflection along the clockwise rotation direction according to an upwind observer. For both deflections, the mismatch with the reference is below 2%.

In fig. 3, the angles describing the flapwise ψ and pitchwise θ blade deflection components are shown for the five flexible cases. Both quantities reach their largest values in the outboard blade, as expected for rotor blades. The flapwise angle increases with wind speed until rated power where its largest value is achieved due to the largest thrust at that condition, as reported

Table 2: Comparison between thrust (Thr) and power (Pow) for the BEMT-based aeroelastic code and the reference results [2] for both rigid and flexible rotor cases at rated conditions.

Case	BEMT Thr	Ref. Thr	Δ Thr	BEMT Pow	Ref. Pow	Δ Pow
Rigid	759.1 kN	761.7 kN	-0.3%	5.56 MW	5.51 MW	0.9%
Flexible	765.3 kN	771.3 kN	-0.8%	5.54 MW	5.49 MW	0.9%

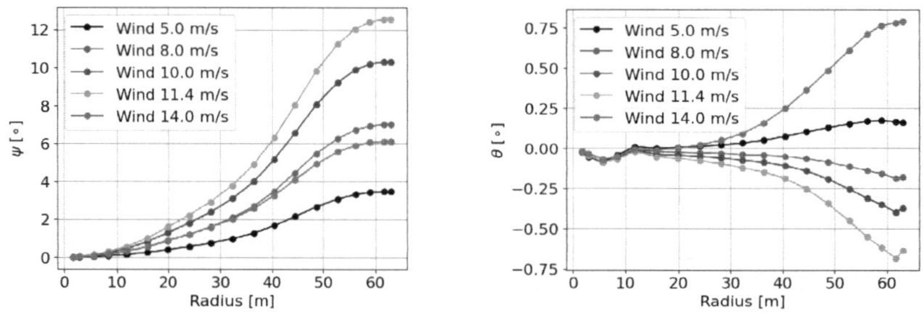

(a) (b)

Figure 3: Angles associated to the (a) flapwise and (b) pitchwise blade deflections predicted by the BEMT code for the simulation cases in table 2.

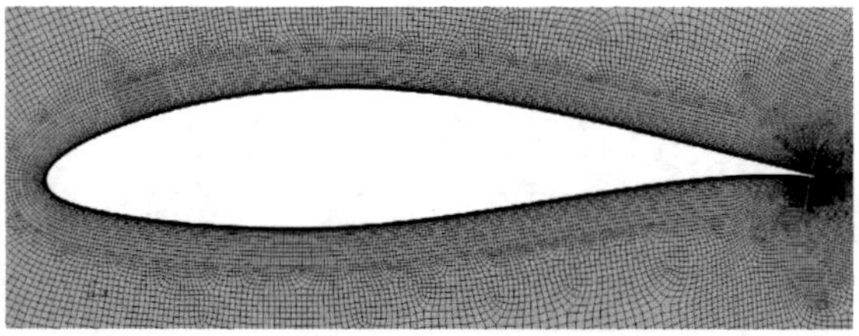

Figure 4: Mesh used for the tip segment airfoil. The pink lines at the trailing-edge correspond to the extraction locations for the boundary layer parameters.

in [7]. The torsion angle follows a similar trend (except for the 5m/s case) and the predicted negative values are related to a "nose-up" rotation of the airfoil section (i.e. opposite to positive aerodynamic twist angle) and increasing the local angle of attack, consistently with Dose's fluid-structure interaction simulations [2].

Finally, the blade structure was also validated by modal analysis of an isolated blade. The frequencies of the first three modes were compared with those reported in [2] obtaining less than 3% difference.

3.2. RANS validation

The five airfoils extracted from the five blade segments share the same set-up for the 2D RANS simulations. The computational domain has the typical C-shape used for airfoil analysis. The velocity inlet boundary condition is imposed in a semi-circumference located 25 chord-unit upstream of the leading-edge of the airfoil. The same boundary condition is imposed at the upper and lower boundaries. The outlet boundary is located 50 chord-unit downstream and a pressure outlet boundary condition is imposed to the same boundary.

The mesh is generated using the automatic quadrilateral mesher of STAR-CCM+. The structured inflation layer is composed by 40 cells in the wall-normal direction, ensuring a thickness of the first cell near the wall of $y^+ \approx 0.3$ and a smooth cell-size transition between the last cell of the inflation layer and the outer unstructured mesh. The resulting mesh for the airfoil used for the tip segment of the blade is shown in fig. 4. Each mesh counts approximately 67 000 elements.

A mesh convergence study has been performed with two additional meshes for each blade segment: a coarser mesh with 47 000 elements and a finer one with 107 000 elements. Both of them have been obtained with a uniform refinement/coarseness of the medium mesh. In

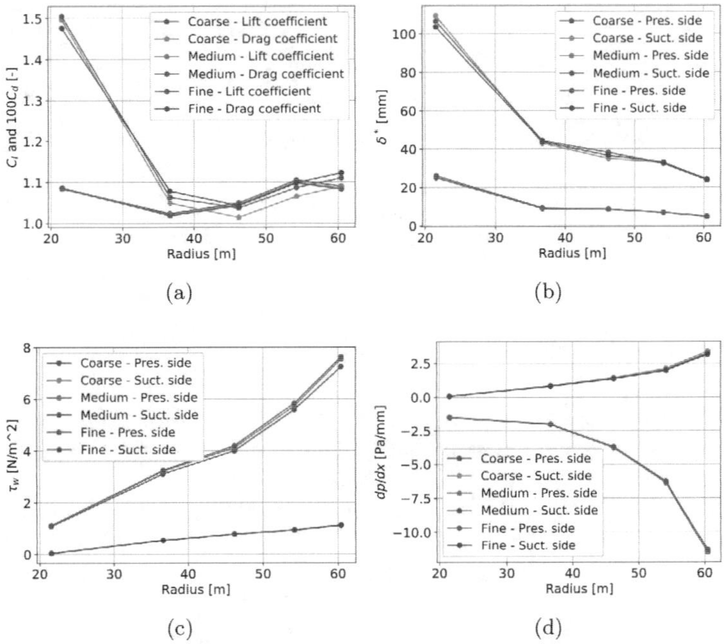

Figure 5: Mesh convergence study showing some parameters of interest as function of the radial position of the airfoil along the blade span for rated operating conditions. (a) Sectional lift and drag coefficients; (b) boundary layer displacement thickness; (c) wall shear stress; (d) pressure gradient. The boundary layer parameters have been extracted at 99% of the chord length.

addition to the lift and drag coefficients, the boundary layer parameters of interest for Amiet's theory are used as indicators for the mesh convergence study. The result shown in fig. 5 for the rated conditions highlights the weak dependence of the mesh utilized on the parameters of interest. Furthermore, for some of them, the values computed with the finer and coarser meshes are almost identical, suggesting a promising robustness with the mesh element size.

As an additional verification for the RANS simulation, the computed lift and drag coefficients are compared to the ones given in [7] and used for the BEMT simulations. Figure 6 compares the RANS results with the BEMT input for both the rigid and flexible rotor cases. The good agreement proves the consistency of the results within the workflow. In fig. 6a, a better agreement is noticed closer to the tip, where the 3D effects not considered in the 2D RANS are negligible. In the flexible rotor case the sectional lift coefficients are higher due to the pitchwise (torsional) deflection. The peak in the lift coefficient for the BEMT input is due to the high angle of attack of the airfoils close to the root and to 3D effects.

Finally, the dependency of the boundary layer parameters on the operating conditions has been used to check the consistency of the trends. As an example, fig. 7 shows the boundary layer displacement thickness for all the operating conditions tested. Up to the rated condition (11.4 m/s of wind speed) the boundary layer thickness on the suction side increases for almost all the blade sections. This is due to the increase in the angle of attack as the wind speed increases. In the above-rated condition (14 m/s wind speed), the pitch control decreases the angle of attack, reducing the boundary layer thicknesses for every blade segment.

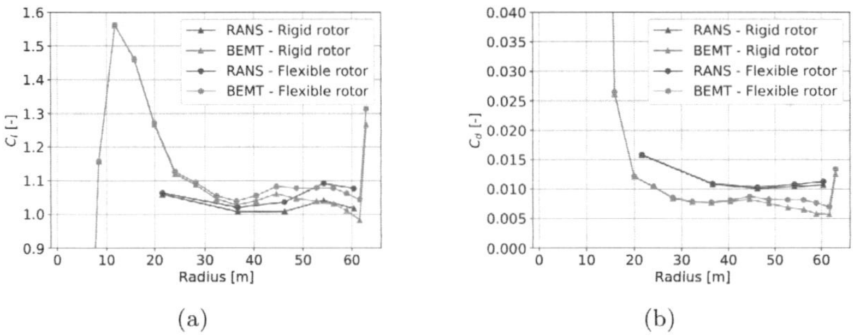

Figure 6: Comparison between RANS results and BEMT input for (a) lift and (b) drag coefficients.

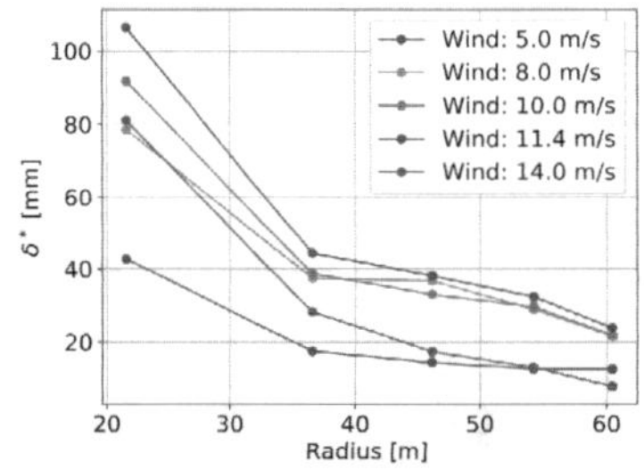

Figure 7: Boundary layer displacement thickness on the suction side of the airfoil for different operating conditions of the wind turbine.

3.3. Amiet's model validation

To the authors knowledge, acoustic measurements or accurate numerical simulations are not available for comparison in current literature. For this reason an in-depth analysis of the trends has been carried out for the simulations with both the rigid and flexible blades.

Figure 8a shows the overall A-weighted sound pressure level computed as

$$\mathrm{OASPL} = 10\log_{10}\frac{\int_{f_1}^{f_2} S_{pp}\mathrm{d}f}{P_0^2}, \tag{3}$$

where $f_1 = 20$ Hz, $f_2 = 20000$ Hz, $P_0 = 2 \cdot 10^{-5}$ Pa and S_{pp} is the A-Weighted power spectral density, sum of the leading- and trailing-edge noise contributions. The power spectral density is computed for an observer located at ground level and 150 m downwind of the wind turbine. The OASPL is represented as a function of the power output computed with the BEMT. Each point on the curves represents an operative condition up to the rated condition. A qualitative comparison can be made with the work of Oerlemans and Schepers [8], where the same trends are reported, supported by experimental measurements.

Figure 8b shows the A-weighted sound pressure levels for the simulation with flexible blades in third-octave bands compared with the results of McBride and Burdisso [45]. In [45] the authors presented the results of NREL 5 MW wind turbine noise simulations using semi-empirical models coupled with an Hamiltonian ray tracing method for the atmospheric propagation. Several differences exist between the approach adopted in the reference [45] and in the present work and, hence, the results must be compared with caution. First, in the present work, the prediction does not include any atmospheric propagation model nor a wind speed gradient, differently

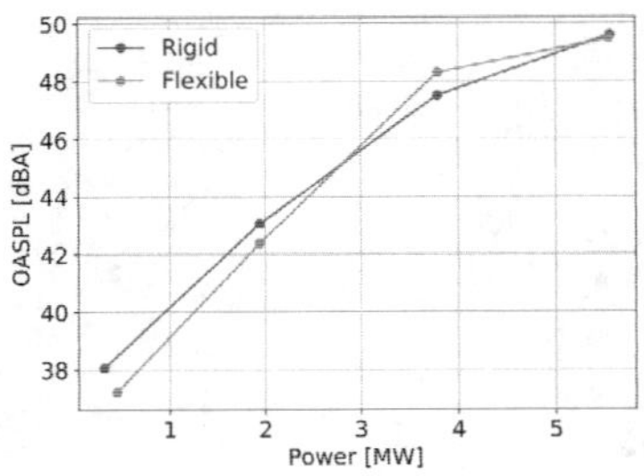

(a)

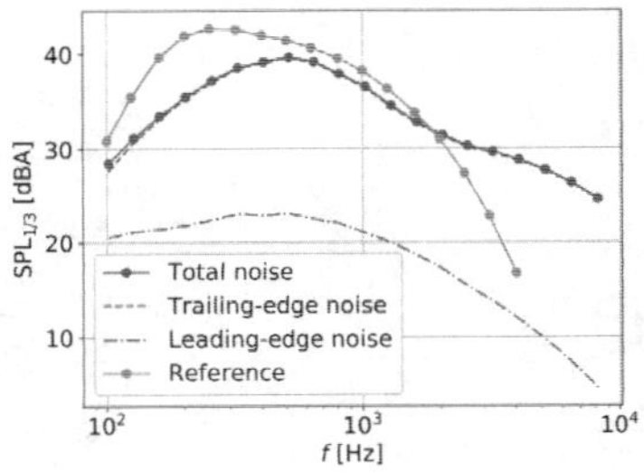

(b)

Figure 8: Validation of Amiet's model. Observer located at ground level, 150 meters downwind. (a) Overall A-weighted sound pressure level. Each point represents a different operating condition up to the rated one. (b) Third octave, A-weighted sound pressure levels compared with results from reference [45] for a wind speed of 10 m/s.

from [45]. Furthermore, in [45], only the trailing-edge noise is computed using the BPM empirical model [46]. Nonetheless, in the mid-frequency range (400 Hz – 2.6 kHz), a satisfactory agreement with reference [45] is reported, while, significant deviations are observed in the low and high frequency ranges. The differences in the low frequency range can be attributed to the different trailing-edge noise models, since it is the main noise source for this test case: as shown in fig. 8b the leading-edge noise is several dB lower than the trailing-edge noise. The differences in the high-frequency range can be attributed to the atmospheric absorption which is expected to significantly affect the high-frequency far-field noise and it is not considered in the present work.

4. Results and discussion

4.1. Influence of the pitchwise and flapwise deflections

To assess the separate effect of the pitchwise and flapwise deflections, an additional acoustic simulation has been performed considering the pitchwise deflection only. The comparison between the noise predictions obtained considering rigid, pitchwise-flexible and fully-flexible blades is shown in fig. 9. The leading- and trailing-edge noise is computed at ground level up to a distance of 5 times the rotor radius. The rotor plane is the XZ plane and the wind is directed as the positive Y axis. Negligible differences are observed between the rigid (fig. 9a) and pitchwise-flexible simulations (fig. 9b), while the effect of the flapwise deflection (fig. 9c) is noticeable for observers close to the rotor plane: the shadow region is much less pronounced when the flapwise deflection is included, resulting in differences up to 13 dBA in the overall sound pressure level. On the contrary, negligible difference are observed in the upwind and downwind directions.

An important finding from the present study is that the symmetry with respect to the rotor plane is lost when the flapwise deflection is considered. For the rigid case the results are almost symmetric as shown, for example, in [8]. This is confirmed by the present work and shown in fig. 9a. On the contrary, the flapwise deflection introduces an asymmetry close to the rotor plane, as it can be observed in fig. 9c. This fact is more evident from the third-octave band analysis of the pressure shown in fig. 10 for the 1 kHz central frequency band. In fig. 10a the directivity pattern resulting from the simulation with rigid blades is symmetrical with respect to the rotor plane, while in fig. 10b higher noise levels are predicted in the downwind direction. The asymmetry is less pronounced for lower frequency bands.

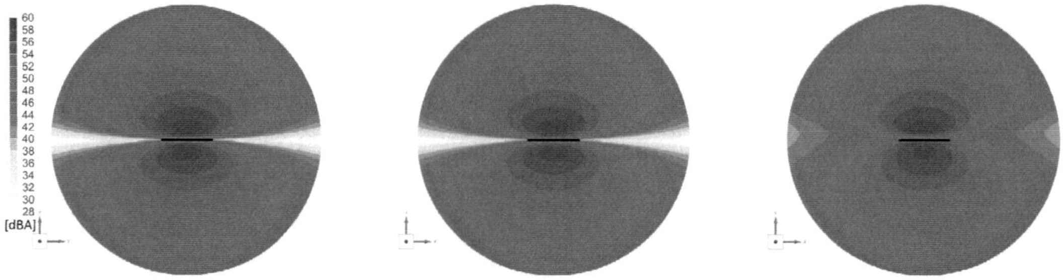

(a) (b) (c)

Figure 9: Footprint of the overall, A-weighted sound pressure level at rated conditions for the (a) rigid-blade simulation, (b) pitchwise-flexible blade simulation, and (c) fully-flexible blade simulation. The thick black line represents the rotor disk.

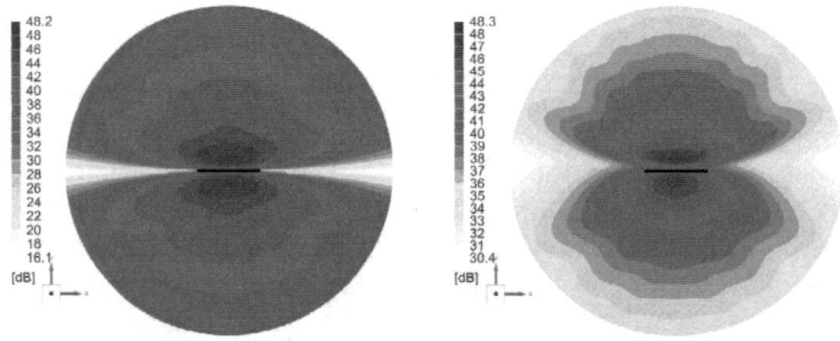

(a) (b)

Figure 10: Footprint of the third-octave pressure for the center frequency 1 kHz. (a) Rigid blades simulation; (b) fully-flexible blades simulation. Note that, differently form fig. 9, the scales are different to highlight the asymmetry in the noise footprints of the flexible blades.

4.2. Leading- and trailing-edge noise

For this wind turbine, the contribution of leading-edge noise to the total noise is negligible at low wind speed. Instead, for the rated and above-rated operating conditions, at low frequency, the leading edge noise becomes the dominant noise source. As an example, fig. 11a shows the A-weighted sound pressure levels for the above-rated conditions at ground level, 150 m downwind. At this observer position, the leading-edge noise becomes the dominant noise source below 100 Hz. This result is also supported by experimental studies from the literature, which state that the leading-edge becomes an important noise source at low frequency for large wind turbines [47]. It is worth mentioning that, even if A-weighted levels at low frequency are smaller than in the medium frequency range, the low frequency noise propagates more efficiently at large distances and, hence, the leading-edge low-frequency noise is expected to have more influence. In fig. 11b, the sound pressure level directivity is shown for the one-third octave band centered at 100 Hz to highlight that the leading edge noise contribution is particularly important in the downwind and crosswind directions.

5. Conclusions

The effect of blade deflections on the NREL wind turbine noise directivity is analysed by coupling the BEMT and a RANS-based Amiet's model. The numerical results show that, when the

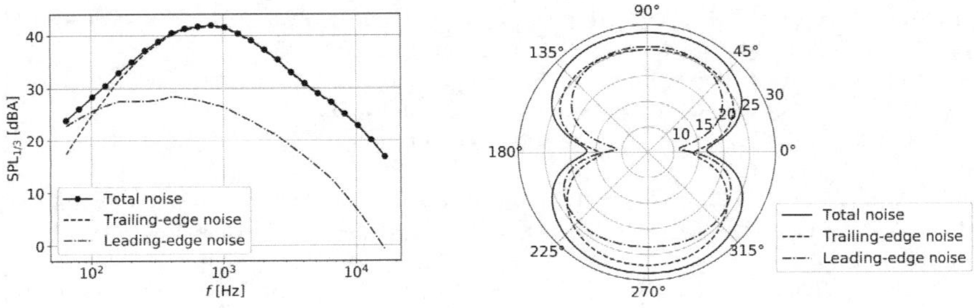

(a) (b)

Figure 11: Noise results obtained with flexible blades and above rated operating conditions. (a) A-weighted sound pressure levels at ground level, 150 m downwind; (b) directivity of the A-weighted sound pressure in [dBA] for the one-third octave center frequency 100 Hz, listeners at ground level and at a distance of 150 m, 90° corresponds to the downwind direction.

flapwise deflection is considered, the shadow region in the rotor plane is less pronounced: in particular, the overall sound pressure level increases in the crosswind direction up to 13 dBA. Furthermore, the symmetry of the results with respect to the rotor plane is lost when the flapwise deflection is included. For these reasons, the implementation of the blade deflections for the noise simulations should be considered for large wind turbines, which undergo significant blade deflections.

For the NREL 5-MW wind turbine considered in this work, the trailing-edge noise is the dominant noise source. The leading-edge noise contribution becomes significant only at high wind speed and low frequency. Furthermore, its contribution to the total noise is relevant in the downwind and crosswind directions.

Acknowledgements

The authors acknowledge the European Commission for its financial support through the H2020-MSCA-ITN-209 project zEPHYR (grant agreement No 860101).

References

[1] Imiela M and Wienke F 2015 Towards multidisciplinary wind turbine design using high-fidelity methods *33rd Wind Energy Symposium*
[2] Dose B, Rahimi H, Herráez I, Stoevesandt B and Peinke J 2018 *Renewable Energy* **129** 591–605
[3] Sayed M, Klein L, Lutz T and Krämer E 2019 *Renewable Energy* **140** 304–318
[4] Rinker J, Gaertner E, Zahle F, Skrzypiński W, Abbas N, Bredmose H, Barter G and Dykes K 2020 *Journal of Physics: Conference Series* **1618** 052052
[5] Kim H, Lee S, Son E, Lee S and Lee S 2012 *Renewable Energy* **42** 46–53
[6] Kaviani H R and Nejat A 2021 *Journal of Wind Engineering and Industrial Aerodynamics* **213** 104617
[7] Jonkman J, Butterfield S, Musial W and Scott G 2009 Definition of a 5-MW reference wind turbine for offshore system development Report National Renewable Energy Laboratory
[8] Oerlemans S and Schepers J G 2009 *International Journal of Aeroacoustics* **8** 555–584
[9] Amiet R K 1975 *Journal of Sound and Vibration* **41** 407 – 420
[10] Amiet R K 1976 *Journal of Sound and Vibration* **47** 387 – 393
[11] Küçükosman Y C, Christophe J and Schram C 2018 *Journal of Wind Engineering and Industrial Aerodynamics* **175** 305–IBC
[12] Küçükosman C 2019 *Semi-analytical approaches for the prediction of the noise produced by ducted wind turbines* Ph.D. thesis TU Delft University
[13] Siemens Digital Industries Software Siemens 2021 Simcenter 3D, version 2021.3

[14] Rozenberg Y, Roger M, Guédel A and Moreau S 2007 Rotating blade self noise: Experimental validation of analytical models *13th AIAA/CEAS Aeroacoustics Conference (28th AIAA Aeroacoustics Conference)*
[15] Tian Y and Cotté B 2016 *Acta Acustica united with Acustica* **102** 626–639
[16] Siemens Digital Industries Software Siemens 2021 Simcenter SAMCEF, version 2022.2
[17] Cardona A 1990 *An integrated approach to mechanism analysis* Thesis Liege University
[18] Hansen M O L, Sørensen J N, Voutsinas S, Sørensen N and Madsen H A 2006 *Progress in Aerospace Sciences* **42** 285–330
[19] Wang L, Liu X and Kolios A 2016 *Renewable and Sustainable Energy Reviews* **64** 195–210
[20] Glauert H 1935 *Airplane Propellers* (Berlin, Heidelberg: Springer Berlin Heidelberg) pp 169–360 ISBN 978-3-642-91487-4
[21] Burton T, Jenkins N, Sharpe D and Bossanyi E 2011 *Wind Energy Handbook* second edition ed (John Wiley and Sons)
[22] Prandtl L 1921 Applications of modern hydrodynamics to aeronautics NACA technical report No. 116
[23] Buhl M L 2005 A new empirical relationship between thrust coefficient and induction factor for the turbulent windmill state Report NREL
[24] Schlinker R H and Amiet R K 1981 Helicopter rotor trailing edge noise Tech. rep. National Aeronautics and Space Administration
[25] Roger M and Moreau S 2004 *AIAA Journal* **42** 536–544
[26] Moreau S and Roger M 2009 *Journal of Sound and Vibration* **323** 397 – 425
[27] Rozenberg Y, Roger M and Moreau S 2010 *AIAA Journal* **48** 951–962
[28] Kücükcoskun K, Christophe J, Schram C and Tournour M 2013 *International Journal of Aeroacoustics* **12** 85–104
[29] Kücükcoskun K, Christophe J and Schram C 2015 *FAN 2015 - International Conference on Fan Noise, Technology and Numerical Methods* **2015-April** 1–12
[30] Bertagnolio F, Madsen H A and Fischer A 2017 *Wind Energy* **20** 1331–1348
[31] Sinayoko S and Hurault J 2015 On predicting wind turbine noise and amplitude modulation using Amiet's theory *6th International Conference on Wind Turbine Noise* pp 1–15
[32] Roger M and Moreau S 2005 *Journal of Sound and Vibration* **286** 477 – 506
[33] Bresciani A P, Le Bras S and de Santana L D 2022 *Journal of Sound and Vibration* **524** 116742
[34] Corcos G M 1964 *Journal of Fluid Mechanics* **18** 353–378
[35] Goody M 2004 *AIAA Journal* **42** 1788–1794
[36] Rozenberg Y, Robert G and Moreau S 2012 *AIAA Journal* **50** 2168–2179
[37] Lee S 2018 *AIAA Journal* **56** 1818–1829
[38] Kamruzzaman M, Bekiropoulos D, Lutz T, Würz W and Krämer E 2015 *International Journal of Aeroacoustics* **14** 833–882
[39] Stalnov O, Chaitanya P and Joseph P F 2016 *Journal of Sound and Vibration* **372** 50–68
[40] Siemens Digital Industries Software Siemens 2021 Simcenter STAR-CCM+, version 2021.3
[41] Siemens Digital Industries Software Siemens 2021 Simcenter STAR-CCM+ User Guide v. 2021.1
[42] Counihan J 1975 *Atmospheric Environment (1967)* **9** 871–905
[43] Zhu W J, Heilskov N, Shen W Z and Sørensen J N 2005 *Journal of Solar Energy Engineering* **127** 517–528
[44] Sinayoko S, Kingan M J and Agarwal A 2013 *Proceedings of the Royal Society A: Mathematical, Physical and Engineering Sciences* **469**
[45] McBride S, Burdisso R and Parra J D 2016 *ICSV 2016 - 23rd International Congress on Sound and Vibration: From Ancient to Modern Acoustics*
[46] Brooks T F, Pope S D and Marcolini M A 1989 Airfoil self-noise and prediction Tech. rep. NASA, Langley Research Center
[47] Buck S, Oerlemans S and Palo S 2016 *Journal of Sound and Vibration* **385** 219–238

Journal of Physics: Conference Series 2257 (2022) 012013 doi:10.1088/1742-6596/2257/1/012013

Investigation of a converter fault for a DFIG wind turbine and analysis of the resulting gearbox component loads

Julian Röder, Georg Jacobs, Dennis Bosse, Fabian Herzog, Lukas Graf

Center for Wind Power Drives (CWD), RWTH University, 52062 Aachen, Germany

julian.roeder@cwd.rwth-aachen.de

Abstract. Doubly fed induction generators are widely used in wind turbines in the field. The main advantage is the significantly lower cost of the partial scale power converter compared to the full scale power converter. Converter faults in wind turbines lead to significant generator torque excitations that lead to dynamic loads in the drivetrain. Dynamic loads and changing rotational speeds can lead to gearbox damages. The generator torque excitations have the highest influence on the high speed shaft torque due to the coupling to the generator. Gearbox damages occur mainly on the components of the high speed shaft. Thus, in this paper the influence of a converter fault in a wind turbine with doubly fed induction generator on the high speed shaft component damages is investigated. It is shown that the converter fault can induce dynamic torque excitations with a maximum increase to around 2.5 times rated torque. Due to the resulting dynamic gearbox loading the safety against scuffing in the high speed gear stage decreases by maximum 19 percent. The smearing risk in the high speed shaft bearings increases to around 2 times the value during rated power production.

1. Introduction

Wind energy will take an increasing share in the energy supply of the future. In order to incentivize the setup of even more wind turbines (WT), costs need to fall further [1]. One driver for WT costs is the reliability. Premature failures of the main components, e.g. gear wheels and bearings, of WTs, directly affect the levelised cost of energy (LCOE) [2]. In order to improve the reliability, all loads during the lifetime need to be considered in the design process. Electrical faults can lead to dynamic generator torque excitations. For WTs with doubly fed induction generator (DFIG) with a partial scale power converter (PSC) grid faults are transmitted directly to the generator which results in dynamic torque excitations with amplitudes significantly higher than the rated torque [3, 4]. Furthermore, power converter faults can lead to highly dynamic torque excitations [5, 6]. These dynamic torque excitations lead to a dynamic loading of the drivetrain components which may result in an increase in the risk of damage to the gearbox components [5]. Damages to the gearbox are the main driver for WT downtime due to the high time to repair the fault. Gearbox damages occur mainly on bearings and gear wheels of the high speed shaft (HSS) [7]. The generator torque excitations due to converter faults have the highest influence on the HSS torque as it is coupled to the generator [8]. Therefore, an investigation of the influence of the torque excitations due to converter faults on the HSS component damages is necessary.

2. Approach

The aim of this paper is to quantify the influence of converter faults on gearbox component damages for DFIG WTs via simulation. In Figure 1 the schematic framework of the WT model used for the load calculation is shown.

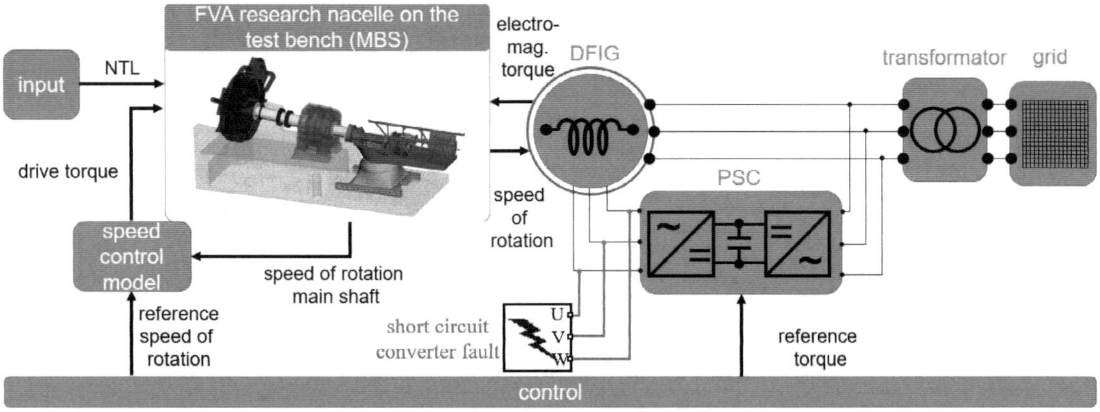

Figure 1. Schematic model of the DFIG research nacelle with the three phase short circuit fault

The model of the WT connects the multi body simulation (MBS) model of the research nacelle drivetrain (see Figure 1), including a detailed model of the gearbox (see Figure 2), with analytical models for the control, the generator, the power electronics and the grid via co-simulation in MATLAB Simulink [9].

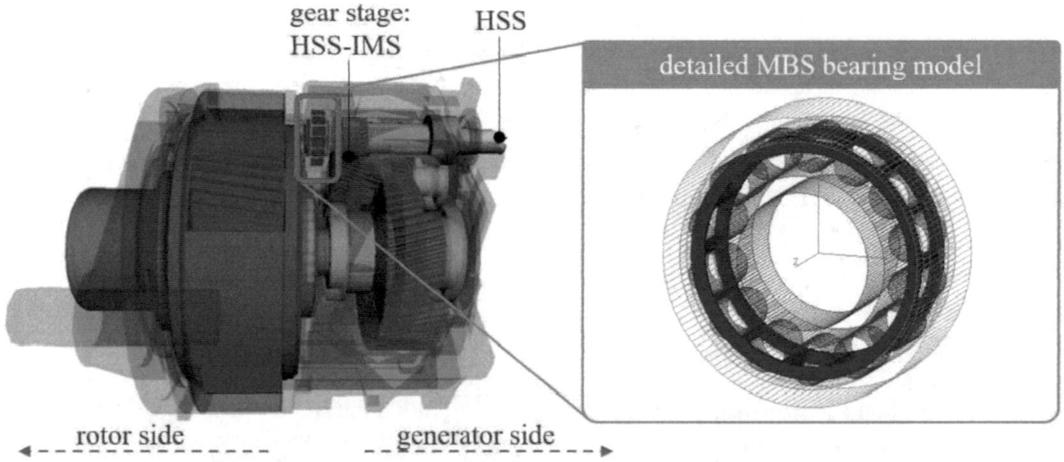

Figure 2. MBS gearbox model with a detailed bearing model

The MBS model also includes the test bench (see Figure 1). The test bench model is used because the same configuration, i.e. DFIG WT nacelle on the test bench, will be set up to conduct extensive experiments regarding electrical faults at CWDs test bench. The test results will enable a validation of the simulation results regarding electrical faults during power production. The model of the research nacelle with the test bench has already been validated via tests of power production according to IEC 61400 disregarding electrical faults [9]. Additionally, experiments regarding a converter fault tested via a crowbar set up with an idling WT were conducted. The tests with the crowbar set up lead to a dynamic torque excitation. Using the measurement data it was possible to validate the dynamic behaviour of the MBS model during the electrical fault for an idling WT [6, 8].

The control model emulates the behaviour of the WT at different operating points (OP). The non-torque loading due to the wind is disregarded in this analysis in order to independently investigate the loading due to the converter fault. The constant vertical loading resulting from the rotor weight is applied. The speed control sets a drive torque, which is applied by the motor, with regard to the actual speed and the reference. Accordingly, the HSS rotates with a rotational speed with regard to the OP until the fault is triggered. After the occurance of the fault, the drive torque is kept constant, which is a sufficient approximation of the torque due to the wind for the short fault effect duration of around 500 ms.

The electrical modelling of the DFIG is comprehensively described in the literature [10-12]. Thus, only the most important aspects are introduced in this paper. Equation (1) shows the stator and rotor voltages of a DFIG in the dq-axis for a synchronous reference frame.

$$\begin{aligned} u_{ds} &= R_s i_{ds} + \frac{d\Psi_{ds}}{dt} - \omega_s \Psi_{qs} \\ u_{qs} &= R_s i_{qs} + \frac{d\Psi_{qs}}{dt} - \omega_s \Psi_{ds} \\ u_{dr} &= R_r i_{dr} + \frac{d\Psi_{dr}}{dt} - (\omega_s - \omega_r)\Psi_{qr} \\ u_{qr} &= R_r i_{qr} + \frac{d\Psi_{qr}}{dt} - (\omega_s - \omega_r)\Psi_{dr} \end{aligned} \tag{1}$$

Here, u is the voltage, R the resistance, i the current, Ψ the flux linkage and ω the rotational frequency. The indices d and q indicate the direct and quadrature axis components and the indices s and r indicate stator and rotor quantities, respectively. Equation (2) describes the slip s of the DFIG:

$$s = \frac{\omega_s - \omega_r}{\omega_s} \tag{2}$$

Consequently, both the frequency of the stator currents, which corresponds to the grid frequency, and the frequency of the rotor currents determine the slip. The speed of rotation of the rotor n_r directly affects the frequency of the rotor currents ω_r, dependent on the number of pole pairs p:

$$\omega_r = n_r \cdot p \tag{3}$$

Subsequently, the slip is dependent on the OP of the WT. For rotor current frequencies above the grid frequency the slip is negative and the machine is in super-synchronous operation mode. Contrary, for rotor current frequencies below the grid frequency the slip is positive and the machine is in sub-synchronous operation mode. Equation (4) shows the electromagnetic torque T_{el} of a DFIG in a synchronous reference frame.

$$T_{el} = \frac{3pL_m}{2}\left(i_{qs} i_{dr} - i_{ds} i_{qr}\right) \tag{4}$$

Here, L_m is the magnetizing inductance and i is the current. The parameters of the generator are implemented according to manufacturer's data (see Table 1).

Table 1.: Main specification of the generator

Specification	Value
Number of pole pairs	3
Apparent power	3 MVA
Rated voltage	720 V
Rated current	2564 A
Rated torque	24.7 kNm

In order to emulate a damage in the machine-side phase module of the converter a three phase fault is implemented between the generator and the converter (see Figure 1). As soon as the short circuit is triggered, the generator of the WT is solely directly connected to the grid. Thus, u_{dr} and u_{qr} approach zero which introduces dynamics in the currents and flux linkage (compare Equation 1). According to Equation 3 this also effects the electromagnetic torque. Additionally, the slip is no longer controlled by the converter but is instead dependent on the characteristics of the generator. It can be assumed that the slip is close to zero for a short circuited rotor [12]. With this assumption, the rotational frequency of the rotor currents approaches the synchronous frequency, i.e. grid frequency (compare Equation 2). In order to achieve the change in speed of rotation there has to be a torque excitation. The resulting dynamic electromagnetic torque excitations are transferred to the gearbox.

The aim of this paper is the quantification of the gearbox damage risk due to these excitations. Focus is on the HSS components because they are the main contributor to WT gearbox damages.

3. Results and discussion

In order to analyse the increase in the risk of damage due to the converter fault, the component load time series need to be evaluated via damage criteria. In this paper the focus is on dynamic driven damages like scuffing for gear wheels and smearing in bearings.

3.1. Torque excitation

The torque excitation for the DFIG WT for the described converter fault is examined for one sub-synchronous and one super-synchronous OP at rated torque of 24.7 kNm. The sub-synchronous OP operates at a speed of rotation of 936 rpm (around 46.8 Hz) (see Figure 3, OP 936). Thus, the generator has a positive slip (compare Equation 2). Therefore, the speed of rotation needs to increase after the fault is triggered. Consequently, the electromagnetic torque decreases rapidly to around -12 kNm before it increases and stabilises at the rated torque after the speed of rotation of around 986 rpm (around 49.3 Hz) is reached. The second OP is at a speed of rotation of 1147 rpm (see Figure 3, OP 1147 rpm), which corresponds to a frequency of 57.3 Hz and therefore results in a negative slip and a super-synchronous operation mode. Therefore, the speed of rotation needs to drop after the fault is triggered. Thus, the electromagnetic torque increases sharply to around 64 kNm (around 2.5 times rated torque). Then it oscillates with amplitudes of around 20 kNm before it stabilises when a speed of rotation of around 1032 rpm (around 51.6 Hz) is reached.

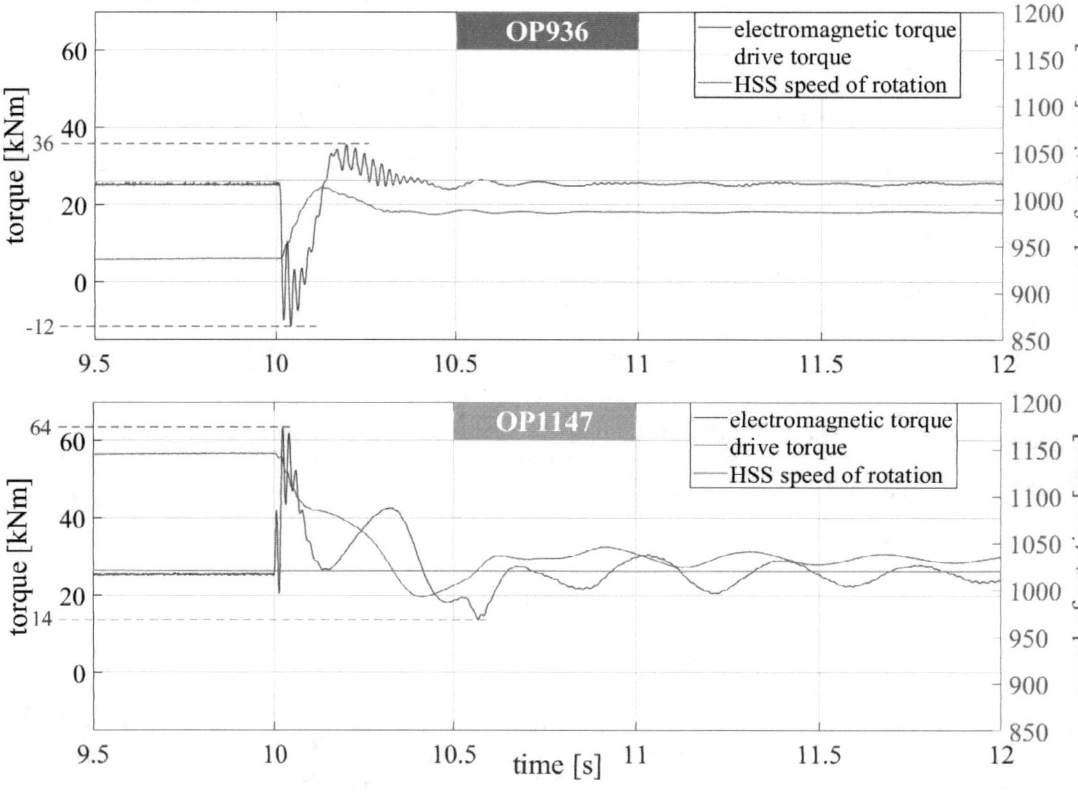

Figure 3. Electromagnetic and drive torque; HSS speed of rotation; DFIG WT with converter fault

3.2. Risk of scuffing damage

For the evaluation of the risk of scuffing damage the contact temperature Θ_B of the meshing gear wheels is calculated according to ISO 6336 [13]:

$$\theta_B = \theta_M + \theta_{fl} \quad (5)$$

Θ_M is the interfacial bulk temperature which can be approximated as the sum of the oil temperature Θ_{Oil}, the max. flash temperature $\Theta_{fl,max}$, the lubrication system factor X_S, which is 0.2 for gears with sufficient cooling and the multiple mating pinion factor X_{mp}, which is 1.0 for the gear stage HSS-IMS [13]:

$$\theta_M = \theta_{Oil} + 0.47 \cdot X_S \cdot X_{mp} \cdot \Theta_{fl,max} \tag{6}$$

Θ_{fl} the flash temperature, which can be calculated using the time series for the HSS torque and speed of rotation as an input for a flash temperature method model in accordance to ISO 6336 [13], that is based on the following formula established by Blok [14]:

$$\theta_{fl} = 2.52 \cdot \frac{\mu_m \cdot X_M \cdot X_J}{50} \cdot \sqrt[4]{(X_\Gamma \cdot w_{Bt})^3} \cdot \sqrt{\frac{n_1}{60}} \cdot \frac{abs\,(\sqrt{\rho_{y1}} - \sqrt{\rho_{y2}/u})}{\sqrt[4]{\rho_{rely}}} \tag{7}$$

where μ_m is the mean coefficient of friction, X_M is the thermo-elastic factor, X_J is the approach factor, X_Γ is the load sharing factor, w_{Bt} is the transverse unit load derived from the tangential force, n_1 is the speed of rotation of the pinion in rpm, ρ_{y1} and ρ_{y2} are the local radii of curvature of the pinion and gear wheel flank while ρ_{rely} is the local relative radius of curvature. Subsequently, the combination of loading, sliding velocity and the resulting contact friction leads to a varying contact temperature along the path of contact [14-16].

In Figure 4 the tangential force in the gear stage HSS-IMS for the OPs is shown. As can be seen, the force oscillates with max. amplitudes of around 21 percent due to the fault, dependent on the OP.

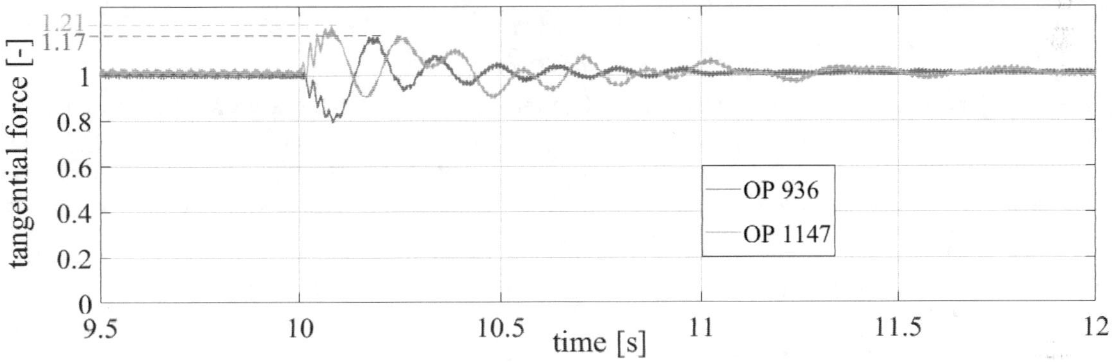

Figure 4. Tang. force of the gear stage HSS-IMS, normalized by the value during nominal power prod.

Furthermore, the angular misalignment of the meshing gear wheels influences the load distribution and the max. loading along the path of tooth contact [17]. The load increase due to misalignment is regarded in Θ_{fl} via the face load factor $K_{H\beta}$ which influences w_{Bt} linearly. For steady nominal power production the gear stage HSS-IMS has an inclination (see Figure 5, left) of around 10.1' and a skew (see Figure 5, left) of around 2.6'. This angular misalignment results in a mean value for $K_{H\beta}$ of around 1.67. Due to the converter fault, $K_{H\beta}$ increases further to a max. of 1.99 (see Figure 5, right).

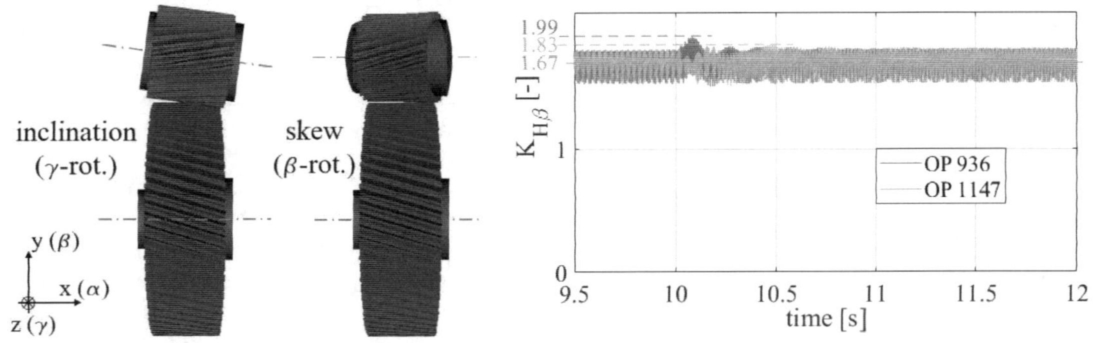

Figure 5. Angular misalignment of the gear stage HSS-IMS (left); $K_{H\beta}$ for the three OPs (right)

In Figure 6 the resulting max. contact temperature $\Theta_{B,max}$ is shown for the two OPs normalized by the mean value of the synchronous operation $\Theta_{B,max,syn}$ of the DFIG WT.

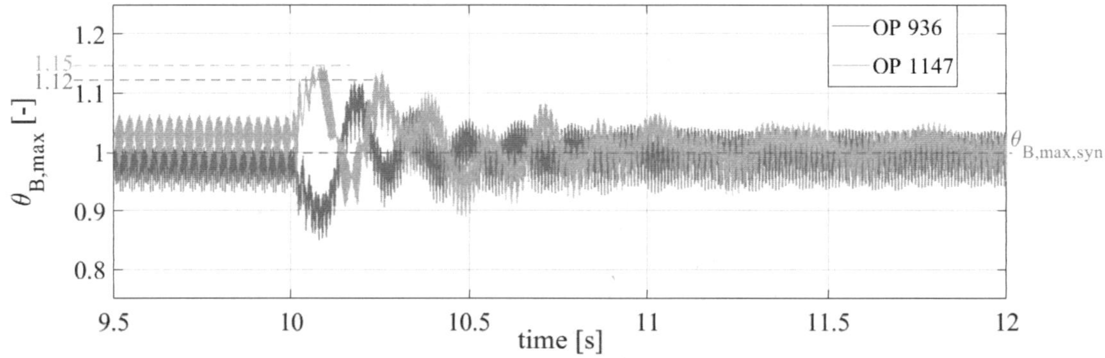

Figure 6. $\theta_{B,max}$ normalized by the mean value of the synchronous operation of the DFIG WT

It can be seen that the overloading, in combination with the changing rotational speed, leads to an increase in $\Theta_{B,max}$ of max. 15 percent for OP 1147 and of 12 percent for OP 936 compared to $\Theta_{B,max,syn}$. In order to quantify the increase in the risk of scuffing due to the rise in contact temperature, the decrease in the safety against scuffing can be calculated via the following formula [13]:

$$S_B = \frac{\theta_S - \theta_{Oil}}{\theta_{B,max} - \theta_{Oil}} \tag{8}$$

where Θ_S is the experimentally determined scuffing temperature. The fault decreases S_B, compared to the mean value of synchronous operation $S_{B,syn}$ by around 17 percent for OP 1147 and by around 19 percent for OP 936 (see Figure 7). Figure 7 also shows that S_B, even after the decrease due to the converter fault, is still well above 1. Consequently, the risk of scuffing damage increases due to the converter fault, but the probability of damage in the shown OPs is still low.

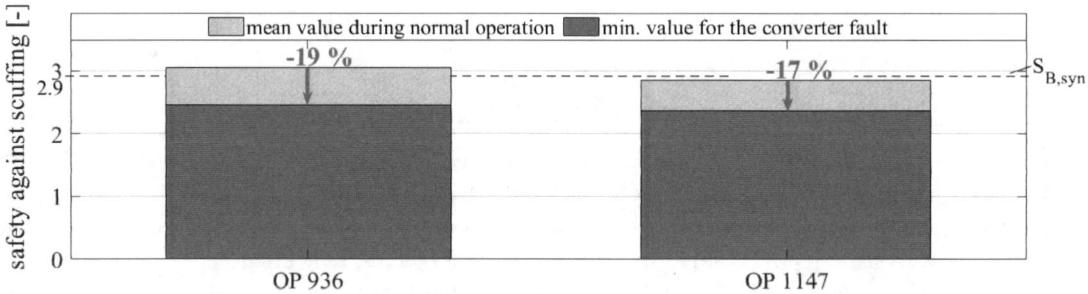

Figure 7. Comparison of the changes of safety against scuffing for the different OP

In order to determine the critical OPs for scuffing in the DFIG WT the heat map shown in Figure 8 can be analysed. On the vertical axis the speed of rotation is shown, while on the horizontal axis the values of w_{Bt} are listed normalized by the mean value of the synchronous operation of the DFIG WT. w_{Bt} is chosen because it combines the loading as well as e.g. misalignment. As can be seen, $S_{B,syn}$ of the DFIG WT is at around 2.9 (also see Figure 7). The heat map shows that for a constant synchronous speed of rotation of around 1000 rpm, w_{Bt} has to more than double in order to decrease S_B to the critical value of 1. For a w_{Bt} that is kept constant at the value of synchronous operation the speed of rotation has to increase to values that far exceed the values of the operating mode of the WT even for extreme load cases in order to reach a critical scuffing state. Therefore, an independent increase in speed of rotation as the cause of scuffing can be ruled out. In conclusion, S_B is only reduced to critical values for extreme load cases with a combined and high increase in speed of rotation and w_{Bt}. For normal power production and for the converter faults at the OPs shown in this paper the safety against scuffing in the DFIG WT gearbox is sufficient.

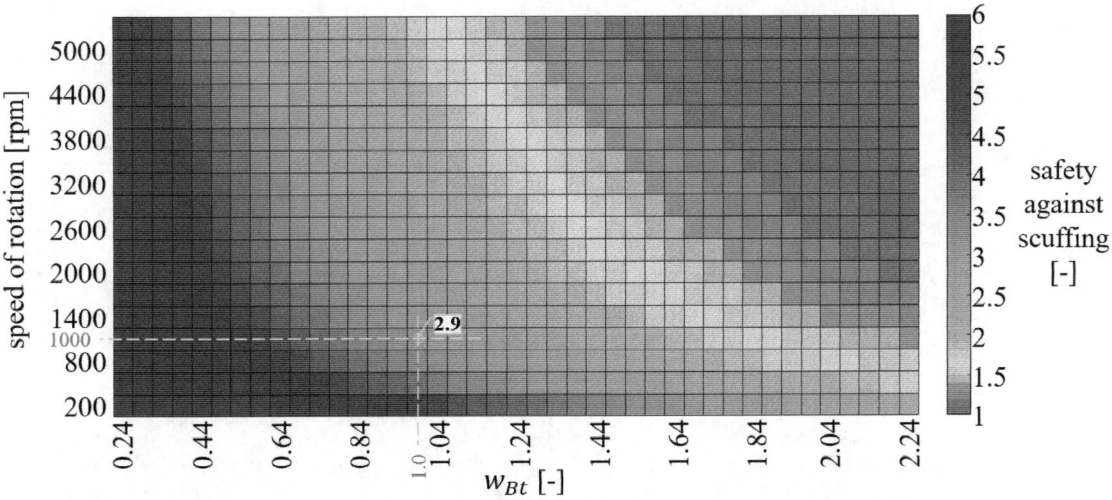

Figure 8. Heat map for the safety against scuffing for the DFIG WT research nacelle

3.3. Risk of smearing damage

In order to calculate the increase in the risk of smearing due to the converter fault the time series for the load and the kinematics of the roller of the HSS cylindrical roller bearing (see Figure 2) are used as an input for a smearing criterion. Smearing damage is triggered by a combination of changing loads and changing speeds of rotation due to an increase in the introduced power P per loaded area A. The smearing criterion used in this paper is shown in the following equation [18]:

$$\left(\frac{P}{A}\right)_{max} = \max\{0.5 \cdot \mu(t) \cdot p_{max}(t) \cdot [u_1(t) - u_2(t)]\} \quad (9)$$

$\mu(t)$ is the friction coefficient, $p_{max}(t)$ is the maximum Hertzian contact pressure and $u_1(t)$-$u_2(t)$ is the difference of circumferential velocity in the contact between the roller and the raceway.

The loading of the HSS bearing is distributed to 14 rollers. In Figure 9, $p_{max}(t)$ for the highest loaded roller for each OP is shown. The increasing HSS loading (see Figure 4) leads to an increase of p_{max} of 8 percent for OP 1147 and 6 percent for OP 936. After the initial increase the loading oscillates before it reaches the rated value again at around 2 s after the fault occurred.

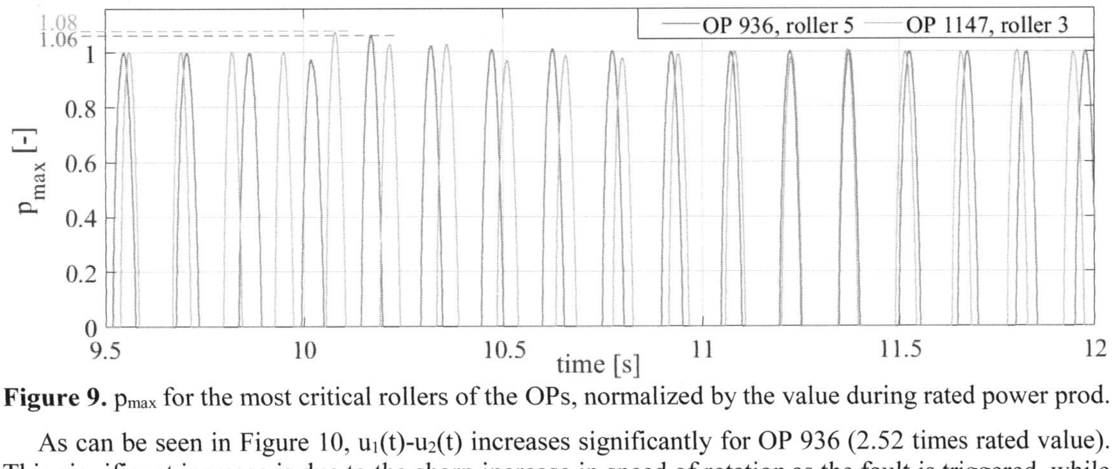

Figure 9. p_{max} for the most critical rollers of the OPs, normalized by the value during rated power prod.

As can be seen in Figure 10, $u_1(t)$-$u_2(t)$ increases significantly for OP 936 (2.52 times rated value). This significant increase is due to the sharp increase in speed of rotation as the fault is triggered, while the generator torque declines (see Figure 3 and Figure 10). Furthermore, due to the fast change in speed of rotation, there is a fast change of the periodicity of the loading (see Figure 9). Therefore, the acceleration phase of the rollers is disturbed for the fault effect duration of the converter fault. This leads to an increase to 1.53 times rated value for OP 1147.

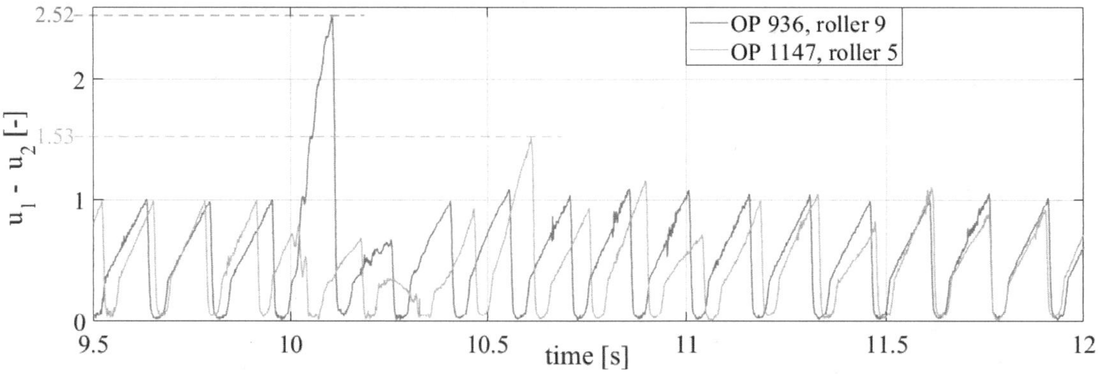

Figure 10. $u_1(t)$-$u_2(t)$ for the most critical rollers of the OPs, norm. by the value during rated power prod.

The critical operating condition for smearing is the combination of high p_{max} while $u_1 - u_2$ is still elevated, which occurs especially in the acceleration phase of the roller. This also means that the most critical roller regarding p_{max} or $u_1 - u_2$ does not necessarily have to be the most critical roller regarding smearing. In Figure 11 the smearing criterion for the most critical rollers for both OPs can be seen. For OP 936 the risk of smearing increases by max. 2.06 times rated value. The increase for OP 1147 to 1.34 times rated value is less severe due to the smaller increases of slippage while the loading is only slightly increased due to the fault.

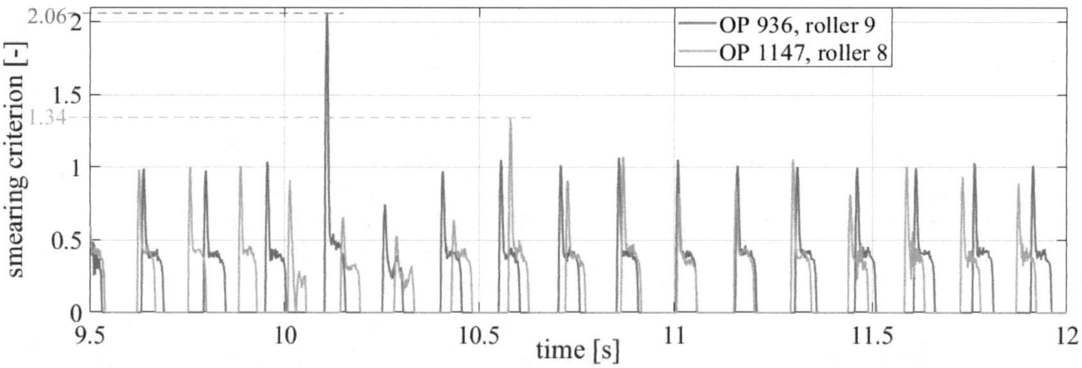

Figure 11. Smearing criterion for the most critical rollers, norm. by the value during rated power prod.

4. Conclusion

A simulative approach for the investigation of the drivetrain loads and the resulting damage risk during a three phase short circuit converter fault for a DFIG WT has been presented in this paper. It was shown that the fault results in a highly dynamic torque excitation, which is dependent on the OP. Due to the resulting dynamic loading in combination with the dynamics in the speed of rotation the risk of dynamic driven damages to the gearbox components is increased. This was shown for the gear wheels via a calculation of the safety against scuffing damage. It was shown that the safety against scuffing damage decreases by max. 19 percent for the analysed OPs. Despite of the decrease, the safety factor for the analysed OPs is still well above 1. Therefore, scuffing damage is unlikely to occur. For the gearbox bearings an evaluation of the risk of smearing damage was done. It was shown that the risk of smearing increases by max. 2 times the value at rated power production during the converter fault.

5. Outlook

The project DynaGET at the CWD will further examine the risk of damage to the components of WT drivetrains during electrical faults via extensive experiments and simulations. The measurements of the research drivetrain at CWD´s testing facility deliver new insights regarding the torque load as well as the loading of the bearings and gears. Using the gathered data, the simulation models will be further validated. The evaluation of the risk of scuffing and smearing will be done for all stages of the gearbox and further fault scenarios. Furthermore, the risk of damage for all currently computable damage mechanisms for bearings and gears of WTs during electrical faults will be evaluated. Knowledge of the damage risk to the gearbox components will enable a determined decision whether countermeasures on the electrical or the mechanical side are necessary for future WTs in order to reduce the WTs downtime due to electrical faults.

6. References

[1] Rohrig, K.: Powering the 21st century by wind energy. Appl. Phys. Rev. 6(3), (2019)

[2] Dhanola A.: Tribological challenges and advancements in wind turbine bearings: A review, Engineering Failure Analysis, Volume 118, doi.org/10.1016/j.engfailanal.2020.104885, (2020)

[3] Erlich I.: Dynamic Behavior of DFIG-Based Wind Turbines during Grid Faults. Power Conversion Conference - Nagoya. IEEE, (2007)

[4] Röder, J.: Investigation of Dynamic Loads in Wind Turbine Drive Trains Due to Grid and Power Converter Faults, In: Energies, 14, 8542. https://doi.org/10.3390/en14248542, (2021)

[5] Röder, J.: Simulative investigation of wind turbine gearbox loads during power converter fault. Forsch. Ingenieurwes. 85, (2021)

[6] Duda, T.: Investigation of dynamic drivetrain behaviour of a wind turbine during a power converter fault. J. Phys. 1037(5), 1456–1463 (2018)

[7] Sheng S.: Wind Turbine Gearbox Reliability Database, Condition Monitoring, and Operation and Maintenance Research Update, Colorado, (2016)

[8] Röder J.: Simulative investigation of the load propagation in a wind turbine drive train during a power converter fault. J. Phys.: Conference series 1618, (2020)

[9] Matzke, D.: Full scale system simulation of a 2.7 MW wind turbine on a system test bench. In: Proc. of conf. for wind power drives, pp. 241–254, Aachen (2017)

[10] Muller, S.: Adjustable speed generators for wind turbines based on doubly-fed induction machines and 4-quadrant igbt converters linked to the rotor. In: Conference Record of the 2000 IEEE Industry Applications Conference, pp. 2249–2254 (2000)

[11] Slootweg, J. G.: Dynamic modelling of a wind turbine with doubly fed induction generator. In: Power Eng. Society Summer Meeting. Conf. Proc. (Cat. No.01CH37262), pp. 644–649 (2001)

[12] Wu, Bin.: Power Conversion and Control of Wind Energy Systems, Wiley, New Jersey (2011)

[13] DIN ISO 6336-20:2019-03, Calculation of load capacity of spur and helical gears – Part 20: Calculation of scuffing load capacity – Flash temperature method (ISO 6336-1:2006)

[14] Blok H.: Measurement of temperature flashes on gear teeth under extreme pressure conditions. In: Proc. Gen. Disc. Lubrication, Inst. Mech. Eng., London, Vol. 2, 14-20, (1937)

[15] Yoon, H.: Scuffing Characteristics of SAE 50B38 Steel Under Lubricated Conditions, In: Tribology Transactions, 45:2, 246-252, DOI: 10.1080/10402000208982547, (2002)

[16] Han, J.: Scuffing behavior of gray iron and 1080 steel in reciprocating and rotational sliding, In: Wear, Volume 271, Issues 9–10, 1854-1861, DOI: 10.1016/j.wear.2010.11.061, (2011)

[17] Theling, J.: Influence of planet carrier misalignments on the operational behaviour of planetary gearboxes. In Proc. of the 8th WZL Gear Conf., Colorado, USA, (2019)

[18] Van Lier, H. Schädlicher Wälzlagerschlupf. FVA Abschlussbericht 1124, Forschungsvorhaben Nr. 663 I; Forschungsvereinigung, Antriebstechnik e.V., Frankfurt, Germany, (2015)

Acknowledgments

The authors thank the Ministry of Economic Affairs, Innovation, Digitalization and Energy of the State of North Rhine-Westphalia, Germany, for the financial support granted. They also thank their project partners for the equipment, insight as well as expertise, which contributed to this joint project.

EUROPÄISCHE UNION
Investition in unsere Zukunft
Europäischer Fonds
für regionale Entwicklung

EFRE.NRW
Investitionen in Wachstum
und Beschäftigung

WindEurope Annual Event 2022 IOP Publishing
Journal of Physics: Conference Series 2257 (2022) 012014 doi:10.1088/1742-6596/2257/1/012014

Lifetime extension and opex reduction by adapting the operational strategy of wind farms

P. Loepelmann and B. Fischer
Fraunhofer Institute for Energy Economics and Energy System Technology (IEE), Königstor 59, 34119 Kassel, Germany

E-mail: peter.loepelmann@iee.fraunhofer.de, boris.fischer@iee.fraunhofer.de

Abstract. The increasing complexity of the requirements for the operation of wind turbines challanges conventional operating strategies. If revenues are generated from the power exchange, it makes sense to react to price fluctuations during operation. How should such responses be designed? What are the implications for the profitability of wind farm projects from an owner's perspective? By carrying out a case study, this paper assesses connecting and disconnecting wind turbines depending on electricity prices using economic criteria. The results of the case study show that it is sensible to shut down turbines in economically unfavorable situations. Lost low revenues are recovered by future revenues over a significantly extended lifetime of the turbines, and the profitability of the project increases considerably.

1. Introduction

Conditions for wind farm operation are changing fundamentally. For instance, new offshore wind farms that are set to begin operation by 2024, are not relying on feed-in tariffs ([1]) and need to generate revenue by directly selling electricity on the power exchange or by securing power purchase agreements. Similarly, wind turbines exceeding twenty years of operation are also losing their feed-in tariff compensation. This will affect wind turbines with an installed capacity of more than 16 GW until the end of 2025 [2]. Additionally, the ongoing energy transition leads to a higher percentage of renewable power in the electric grid, thus intesifying the coupled dynamics of solar and wind power feeds and their market value, respectively [3].

How are these changes in market conditions affecting wind farm operation? In general, wind farms are no longer operated under the premise of maximizing energy production during a certified and fixed turbine lifetime that is primarily related to damage accumulation. In the future, this simple operational strategy has to replaced by more sophisticated schemes in order to take full advantage of the economic potential that is found in optimizing the complex interplay between varying prices, varying production costs, corresponding loads and wear and resulting in a non-fixed turbine lifetime.

This article is focused on the latter aspect. Specifically, we take a look at the economic ramifications of extending lifetime by judiciously powering down turbines. We adress the question of whether deactivation during economically unfavorable conditions and thus attaining extended lifetime will increase the overall wind farm profitability. Conditions that are particularly unfavorable include situations of low power exchange prices and simulataneous high-load wind conditions. These circumstances are quite probable, since wind power prices plummet at most sites during strong winds since lots of wind farms are producing at capacity simultaneously.

In order to investigate these questions, different operational schemes are implemented that engage and disengage wind turbines depending on electricity prices. This significantly extends current schemes that only look at turbine-specific cut-in and cut-out wind speeds. We evaluate the operational schemes using a case study of an existing wind farm in order to assess the economical parameters involved. We assume that a wind farm has been operating under a feed-in tariff scenario for 20 years, after which the power is sold directly at the power exchange at market conditions. The economic evaluation is based on traditional pre-investment analysis: the annuity method.

2. Design problem and model specification

Fig. 1 shows a schematic of the design problem. We are looking for the operational scheme $F(p)$ that automatically connects and disconnects the farm to the grid. Every 10 min the operational scheme decides on the basis of current market price p if the farm operates in power mode or not. This operational scheme should maximize the economical benefit of the owner of the farm, which is represented by the net present value K_N, see Section 2.3. To this end, the operational expenses and the revenues are being discounted and accumulated over the lifetime N of the project. The latter is not fixed but depends on the turbines' technical lifetime that is determined by the fatigue damage with respect to the operation using $F(p)$. The remainder of this section explains the modeling of the subsystems.

Figure 1 suggests that time series are used to investigate the influence of the operational scheme $F(p)$ on the overall economic efficiency. In contrast, the annual values of revenues, costs, and damages are calculated using a probabilistic model.

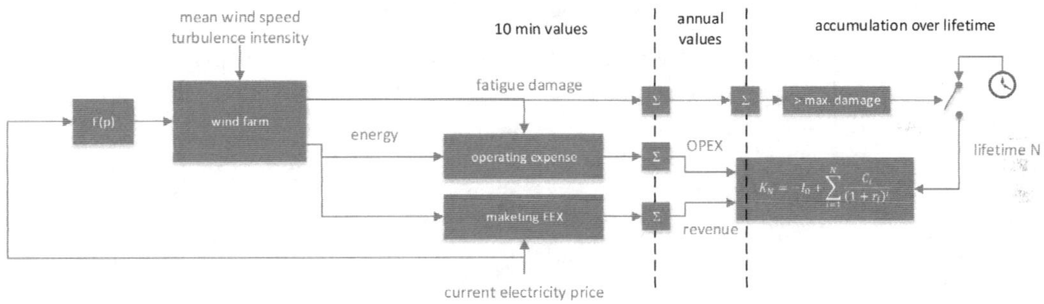

Figure 1. Schematic of the design problem.

2.1. Probabilistic model of wind speed and market price

The probabilistic model represents a joint distribution of environmental parameters and market price p. The environmental parameters are the wind speed v and the turbulence intensity TI with respect to 10 min time frames. Both parameters characterize the most important influences on power output and fatigue loads. The wind speed v is chosen to be the only independent variable to calculate the joint distributions of turbulence intensity TI and market price p.

Figure 2 shows the specific model that was used in the case study. The data represent a potential site in northern Germany near the danish border. Frequency distribution of wind speed v is obtained from a historical time series of the year 2018 from the regional reanalysis REA6 [4] provided by the German Meteorological Service and interpolated to hub height. The conditional distribution of the market price is calculated using the time series of the same year from the European power exchange (EPEX SPOT). The calculation of the conditional distribution of turbulence intensity is based on an IEC-class A type (see [5]).

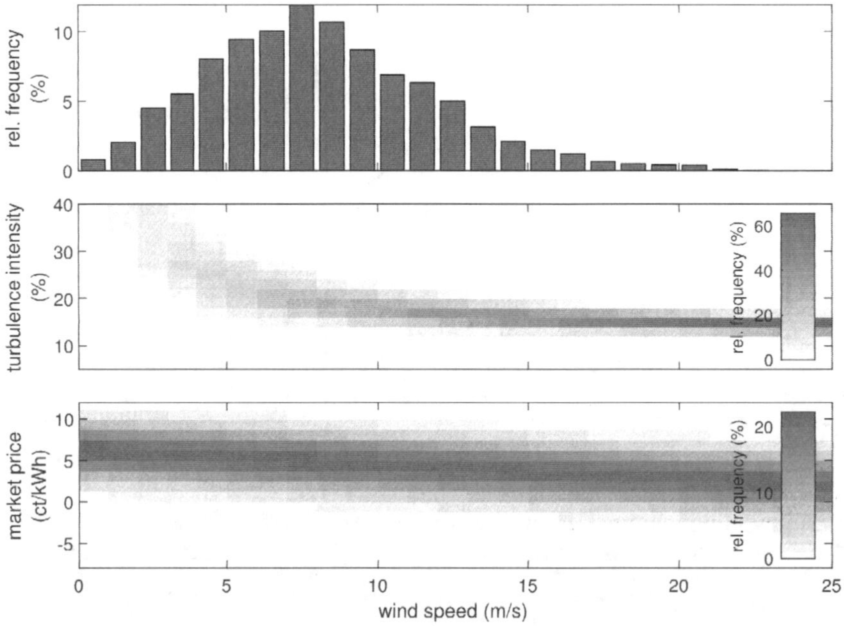

Figure 2. Frequency distribution of wind speed parameters and market price. Top: Frequency of wind speed. Middle: Relative frequency of turbulence intensity. Bottom: Relative frequency of market price.

2.2. Model of wind farm

A surrogate model is used to represent the wind farm in Fig. 1. It is a static, characteristic field that maps the environmental parameters v and TI to the turbine's energy yield E and fatigue damage D of components during 10 min operating time. The surrogate model is parametrized using aero-elastic simulation codes.

In this study, a simplifying assumption is made regarding the fatigue modelling. Only the damage during normal operation is considered. Other damage contributions, e.g. during idling or because of start-up and shut-down maneuvers, are neglected. This assumption does not fundamentally affect the results of the case study.

Fig. 3 exemplifies the normalized ratio D/E as a function of v and TI. The most unfavorable operating situations occur when high wind speeds are combined with the highest turbulence intensities possible. Energy that is generated in this manner is particularly "expensive" in terms of consumption of the technical lifetime of the turbine.

The wind farm outputs D and E, calculated using the surrogate model, are weighted by the probability of co-occurrence of the three variables for all combinations of (v, TI, p) at which the turbine injects energy according to the schedule $F(p)$ and summed over one year. The resulting annual values are used by the investment appraisal approach described in the next section to evaluate profitability over the project's lifetime N.

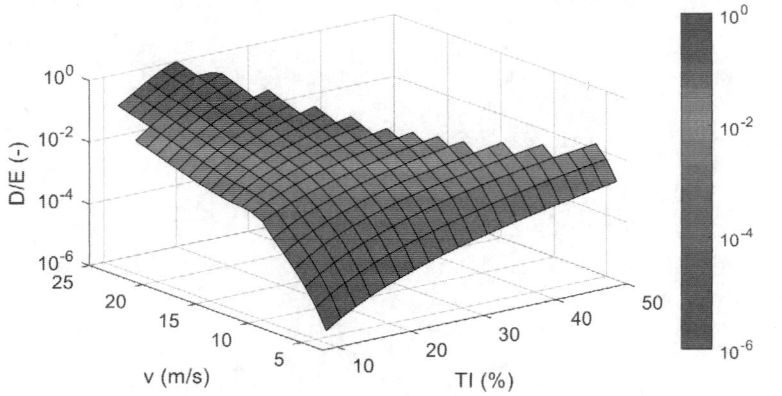

Figure 3. Example of the normalized ratio D/E as a function of v and TI. The ratio D/E is normalised with its maximum.

2.3. Economical assessment

In order to evaluate the economic implications of individual operational strategies, basic models of dynamic pre-investment analysis are used (see [6]). First, the net present value of the investment is obtained for different running times $N \in \mathbb{N}$:

$$K_N = -I_0 + \sum_{i=1}^{N} \frac{C_i}{(1+r_i)^i}, \tag{1}$$

with N denoting the investment running time in years. In (1), annual cashflow $C_i = R_i - O_i$ is the difference of market revenue R_i and operational expenditure O_i. It is subjected to the interest rate r_i and aggregated over the running time. To obtain the net present value, the initially invested capital I_0 for the wind farm is subtracted. If the net present value is positive, the investment returns more revenue than if the initial capital would have been invested given the interest rate. The running time N is either chosen by the investor or determined by the technical turbine lifetime, which is particularly constrained in this study by the tower fatigue damage D_{twr}:

$$N \to \max, \text{ so that } \sum_{i=1}^{N} D_{twr,i} < \bar{D}_{twr}, \tag{2}$$

which means that if the accumulated damage $D_{twr,i}$ exceeds the damage design limit $\bar{D}_{twr}$, the tower fails.

Since net present values of different running times are not comparable, they are converted into annuity values:

$$A_N = K_N \cdot a_N \tag{3}$$

The annuity factor a_N is dependent on the investment running time and interest rate (see [6]). An investment with a higher annuity is favorable to one with lower annuity.

The annual revenues R_i are the sum of revenues $E(v, TI, p) \cdot p$ generated during the different environmental and price conditions (v, TI, p) weighted with their probability of occurrence $P(v, TI, p)$, i.e.

$$R_i = M \cdot \sum_{(v,TI,p)} P(v, TI, p) \cdot E(v, TI, p) \cdot p,$$

where M denotes the number of time intervals during one year. Note that $E(v, TI, p)$ becomes price-dependent due to the operational scheme $F(p)$.

A non-standard approach is used for calculating the annual operational expenditure O_i. The purpose of the particular OPEX model is to reflect cost changes due to different operational schemes. It is assumed that a flexible maintenance contract has been arranged that can account for the actual operation of the turbines – similar to a pay-as-you-drive car insurance. The model is a combination of the reliability model from [7] and the cost analysis in [8]. It features three parts:

$$O_i = O_{fix}(P_{rated}) + O_{var}(E_i, \gamma_i) + O_{rep,i}.$$

The fixed costs $O_{fix} = \alpha \cdot P_{rated}$ are proportional to the rated power of the farm P_{rated}. Variable costs $O_{var}(E_{\Sigma,i}, \gamma_i) = E_i \cdot (\beta + \gamma_i)$ are modeled taking into account the annual energy yield $E_i = M \cdot \sum_{(v,TI,p)} P(v, TI, p) \cdot E(v, TI, p)$ and a parameter γ_i that represents the normalized, annual failure rate, which increases with respect to time and accumulated fatigue damage over the project period. Eventually, the replacement costs $O_{rep,i}$ comprise costs caused by replaceable components, e.g. the converter system. These costs are triggered component-wise in case the component's accumulated failure rate exceeds a certain threshold.

2.4. Additional parameters of the case study

For the case study presented in this article, we define a fictional wind farm consisting of 20 wind turbines of a reference design, namely the TUM-3.35 turbine provided by Technical University of Munich (see [9]). It is a modern turbine design for sites with low mean wind speeds of IEC-class III-A with

- $P_{WEA} = 3.35\,\text{MW}$ rated power,
- cut-in and cut-out wind speeds of 4 m/s and 25 m/s, respectively, and
- a hub height of 110 m.

Loads and damages of the turbines have been changed slightly compared to the design parameters such that they would reach a lifetime of 25 years following continuous operation under a fixed feed-in tariff scheme. If turbines are disengaged for certain stretches during their lifetime - such as during negative electricity prices - longer lifetimes are achieved.

The investment cost of the wind farm I_0 has been adapted to obtain a scenario in which the operation with fixed feed-in tariff is narrowly profitable after 20 years of operation, see Fig. 4. To determine interest rate, we choose (see [8]):

- For the equity capital of 17 %: 8 %
- For debt capital 2.3 % for the first ten years and 5.1 % after that.

Thus, the comprised interest rate r_i in (1) amounts to 3.27 % for $i \leq 10$ and 5.51 % after. For the OPEX model we derived $\alpha = 25\,€/\text{kW}$ and $\beta = 0.004\ €/\text{kWh}$ from [7] and [8].

3. Price-dependent operational scheme

A simple operational strategy is to follow a basic rule: wind turbines are only in power production mode if electricity price p exceeds a minimum value p_{min}:

$$F_p(p) = H(p - p_{min}), \tag{4}$$

where $H(\cdot)$ denotes the Heaviside-function. Iteratively increasing the minimum value yields the results given in Table 1. Higher minimum prices achieve longer lifetimes, since turbines are shut down more often.

Table 1. Annuities of simple operational strategy $F_p(p)$. The parameter p_{min} sets the cut-in price from where the turbines are put into power production mode.

p_{min}/(ct/kWh)	Annuity A_N / M€	Obtained in year
any	0.19	25
0.0	0.28	26
0.5	0.30	26
1.0	0.33	26
1.5	0.35	26
2.0	0.38	27
2.5	0.41	28
3.0	0.43	30
3.5	0.43	32
4.0	0.39	36
4.5	0.28	42
5.0	0.10	50

First, annuities and project lifetimes increase with increased minimum prices. However, very high minimum values, beginning from 4.0 ct/kWh, result in lower annuities, while project lifetimes are further increasing up to utterly unusual values. This annuity reduction is due to reduced annual cashflow that is subjected to interest. The highest annuity of 0.43 M€ is achieved at $p_{min} = \{3, 3.5\}$ ct/kWh with a total lifetime of 30 and 32 years, respectively.
Figure 4 gives a detailed plot of the annuity development of two minimum price settings: 0 and 3.0 ct/kWh. The annuity plot for the wind farm operation from year 0 to year 20 is determined by the fixed feed-in tariff revenue. An annuity of 0 € is reached at the end of this period, so the investment is profitable at this point already. The remaining component lifetime from year 20 on may be utilized by implementing the proposed operational strategy.

3.1. Operation with $p_{min} = 0$ ct/kWh

Consider the case of $p_{min} = 0$ ct/kWh when turbines are only in power production mode if prices are positive. This denotes an obvious operational scheme during market price operation because it is obviously sensible to avoid selling power during negative prices. In our model, this already attains a lifetime extension of one year as compared to the case with conventional strategies for AEP maximization, see Section 2.4. However, no economic advantages are achieved, since the annuity of the investment does not increase from year 25 to year 26, when lifetime damage limit is reached. Nevertheless, the advantages of the devised operational scheme are already observable.

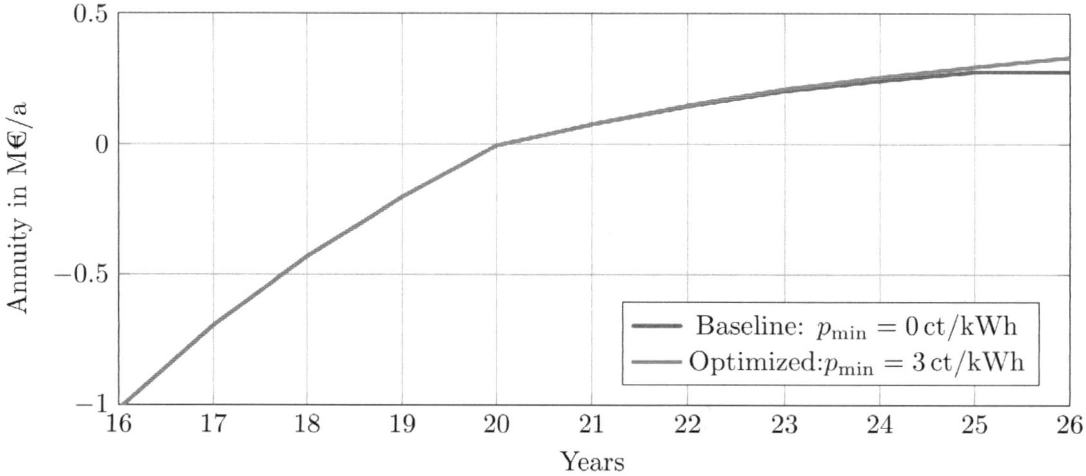

Figure 4. Plot of annuity development for different operational strategies. From 0 to 20 years, annuities are identical, since this period is governed by feed-in tariff operation. After 20 years, the optimized strategy (green) achieves higher annuity than the baseline strategy (red).

3.2. Optimized operation with $p_{min} = 3.0$ ct/kWh

If the results of the annuity calculations in Table 1 are consulted, an optimal minimum price of $p_{min} = 3.0\,\mathrm{ct/kWh}$ is selected. It allows for the maximization of annuity, i.e. most profitable investment, after a running time of 30 years. Considering the annual annuity plot in Fig. 4, it is evident that the operation above a minimum price of 3 ct/kWh achieves both a higher yearly annuity and a much longer technical lifetime and running time of the investment, during which more revenue is created.

3.3. Cashflow analysis

In order to take a closer look at the underlying economic dynamics, cost and revenue plots are depicted in the following figures. First, consider power revenues (red lines) in both Figs. 5 and 6. For the first 20 years, i.e. during feed-in tariff compensation, revenues are high and nearly constant. Higher shut-down times in later years due to component wear lead to a slight decrease in revenue over time. However, after 20 years, power revenues plummet in the absence of feed-in tariffs. Annual revenue is higher while operating the turbines over the minimum price of 0 ct/kWh. However, lower annual revenues in Fig. 6 are achieved over a longer period.
Operational expenditure (green lines) increases continuously when implementing the operational strategy with a minimum price of 0 ct/kWh (see Fig. 5). However, when choosing $p_{min} = 3\,\mathrm{ct/kWh}$, operational expenses are reduced after 20 years, thus contributing to cashflow (see Fig. 6). This is due to the more gentle mode of operation. Thus, cashflow is higher and achieved over a longer time period when choosing the operational strategy based on $p_{min} = 3\,\mathrm{ct/kWh}$.

4. Conclusion

The results of the case study presented in this article demonstrate the possibility of significantly increasing wind farm profitability by disconnecting wind turbines during low-revenue time periods. This is achieved by saving component damage which adds to the overall lifetime of the turbines and reduces operational expenditure.
After a period of 20 years, during which feed-in tariff revenues are generated, the turbines are not at their lifetime damage limit, so continued operation is possible. Optimizing the control

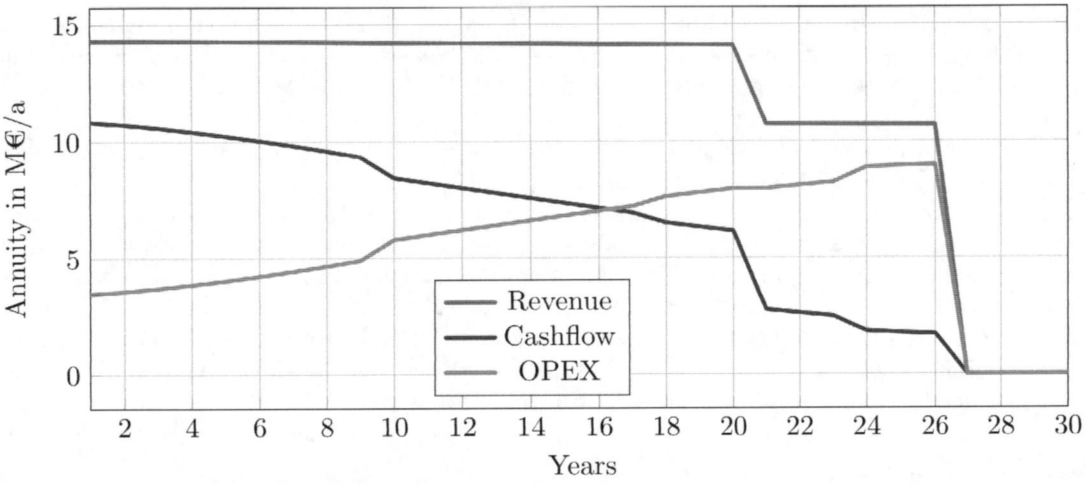

Figure 5. Plot of electricity revenue (red), operational expentidure (green) and resulting cashflow (blue) over the investment running time. Operational strategy with minimum price of 0 ct/kWh.

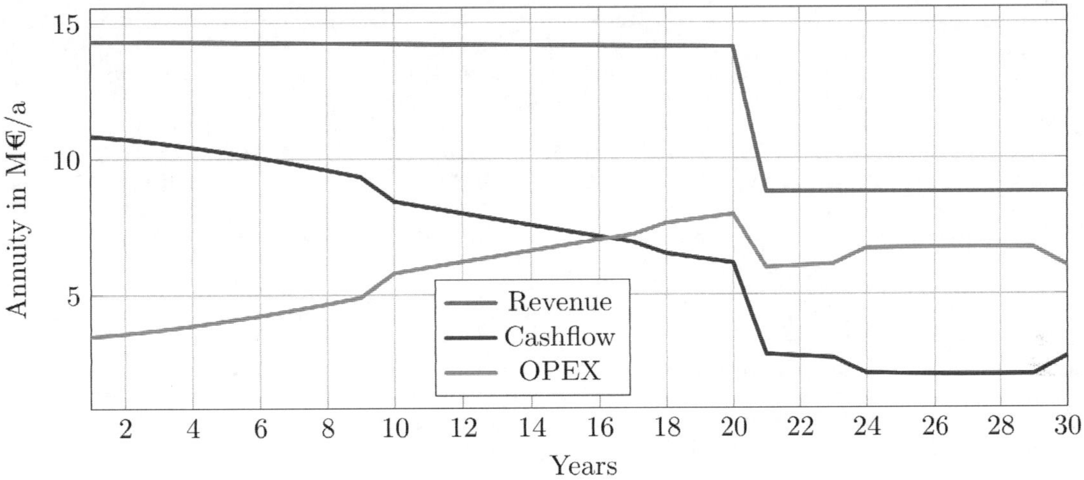

Figure 6. Plot of electricity revenue (red), operational expentidure (green) and resulting cashflow (blue) over the investment running time. Operational strategy with minimum price of 3 ct/kWh.

strategy during this second period, we can extend the technical lifetime and running time of the investment to maximize annuity. This results in higher financial profits as well as increased renewable energy production. In [10] it is shown that profitability can be further increased by choosing a strategy that depends additionally on the environmental conditions, i.e. $F(p, v, TI)$. The profit increase, however, comes at the price of considerably larger project lifetimes and a severely higher implementation complexity.
In any case, longer investment running times are necessary to realize the economic advantages of this approach. Thus, uncertainty regarding interest rates and power exchange prices are increased. Moreover, when trading on a day-ahead market is considered, the influence of prediction uncertainty should be taken into account. The economic modeling of this approach, including the OPEX model, is the focus of further research in this area.

The results of this study are generated with a specific choice of site and wind turbine model. It is thus worthwhile to extend this research to other wind farm examples and to conduct sensitivity analyzes and to investigate parameter variations.
Furthermore, the approach may be extended beyond a simple connect/disconnect decision in the future. It is conceivable to implement power throttling into the optimization which would allow for a much finer tuning of the load-lifetime-revenue dynamics.
There exists another interesting field of research in this context: power exchange prices may be subject to feedback from this operational strategy. It is not obvious how the dynamics between wind conditions and power exchange prices may change if the strategy presented here is implemented by many wind farms. Possibly, the strategy will need to be amended.
Also, the proposed profitability improvement and accompanied lifetime extension competes with a possible repowering of the wind farm. In future work, we will consider how this investment decision is made in light of the presented optimized strategy (see also [11]).

Acknowledgments
This article is based on a research project funded by the Federal Ministry of Economic Affairs and Climate Action under grant 03EE3005A.

References
[1] Bundesnetzagentur. WindSeeG - 1. Ausschreibung für bestehende Projekte nach § 26 WindSeeG, Ergebnisse der 1. Ausschreibung vom 01.04.2017: BK6-17-001; 2017. Available from: `https://www.bundesnetzagentur.de/DE/Beschlusskammern/1_GZ/BK6-GZ/2017/BK6-17-001/Ergebnisse_erste_Ausschreibung.pdf?__blob=publicationFile&v=3.`
[2] J Quentin, D Sudhaus, M Endell. Fachagentur Windenergie an Land, editor. Windenergieanlagen - Was tun nach 20 Jahren?. Berlin;. Broschüe FA Wind, Berlin 2018. Available from: `https://www.fachagentur-windenergie.de/fileadmin/files/Veroeffentlichungen/FA_Wind_Was_tun_mit_WEA_nach_20Jahren.pdf.`
[3] Marc Anders. Analyse von Einflussgrößen auf empirische Marktwerte von Windenergieanlagen in Deutschland [Masterarbeit]. Universität Kassel; 2020.
[4] Bollmeyer C, Keller JD, Ohlwein C, Wahl S, Crewell S, Friederichs P, et al. Towards a high-resolution regional reanalysis for the European CORDEX domain. Quarterly Journal of the Royal Meteorological Society. 2015;141(686):1-15.
[5] International Electrotechnical Commission. Wind energy generation systems - Part 1: Design requirements: IEC 61400-1; 2019.
[6] Petersen P, Fabozzi F. Capital Budgeting: Theory and Practice. Wiley; 2002.
[7] Wiggert M, Meyer-Lerbs L, Wolken-Möhlmann G, Dietrich E. Offshore TIMES - Offshore Transport, Inspection and Maintenance Software : Abschlussbericht;.
[8] A -K Wallasch, S Lüers, K Rehfeldt, K Vogelsang, H Jachmann. Deutsche Windguard, editor. Vorbereitung und Begleitung bei der Erstellung eines Erfahrungsberichts gemäß § 97 Erneuerbare-Energien-Gesetz: Teilvorhaben II e): Wind an Land;. Wiss. Bericht, Deutsche WindGuard, 2019.
[9] Bortolotti P, Tarres HC, Dykes KL, Merz K, Sethuraman L, Verelst D, et al.. IEA Wind TCP Task 37: Systems Engineering in Wind Energy - WP2.1 Reference Wind Turbines;. Bericht NREL/TP-5000-73492, 2019.
[10] Fischer B. Fahrplanoptimierung von Windenergieanlagen ohne feste Einspeisevergütung. at - Automatisierungstechnik. 2022;70(3):292-9.
[11] Piel JH, Stetter C, Heumann M, Westbomke M, Breitner MH. Lifetime Extension, Repowering or Decommissioning? Decision Support for Operators of Ageing Wind Turbines. Journal of Physics: Conference Series. 2019;1222:012033.

Machine learning based met data anomaly labelling

Ana P. Talayero[1], Nurseda Y. Yürüşen[1], Francisco José Sánchez Ramos and Roberto Lázaro Gastón

Fundación CIRCE, Edificio 3D, Primera Planta, Av. de Ranillas, 50018 Zaragoza

E-mail: aptalayero@fcirce.es

Abstract. Data preprocessing is the initial and utmost critical phase of wind resource assessment and wind power curve performance analysis. Without ensuring high quality and site representative data availability, an evaluation for the wind resource potential of a wind farm site means investment decision making involving a great deal of uncertainty. The current practice in the field is to use fix rules and via data scanning by manual labouring of data field experts. Although rule-based (if-else, interval-based) prefiltering applications can be found in some commercial software, and these rules are also limited for addressing the needed preprocessing requirements fully. Therefore, this process is time consuming and causes inefficient resource allocation considering the manual scanning process performed by data scientists & experts. In this study, reviewed data by experts are used as data with correct labels, and then efficient classifiers are generated by applying machine learning algorithms to unfiltered and filtered data. It is reported that tree-based classifiers are performing better.

1. Introduction

In the standard IEC 61400-12-1:2017 [1], the data filtering process is referred to in a very abstract way. Although rule-based (if-else, interval-based) prefiltering applications can be found in some commercial software. These rules are also limited to addressing the needed preprocessing requirements fully. Therefore, this process is time-consuming and causes inefficient resource allocation considering the manual scanning process performed by data scientists & experts. There are weather sensor fault & outliers detection studies, which focus on developing algorithms for precipitation and temperature data anomalies [2]. In wind energy literature, machine learning based solutions can be found in different sub-fields and their data preprocessing applications, as an example in wind turbine condition monitoring [3] and in wind energy forecasting [4]. Nevertheless, the machine learning based meteorological data filtering and the comparison of findings with the industry practices are found implicit and scarce in the literature. As an example, this problem is treated as an anomaly detection study [5] previously; nevertheless, it has a distinct nature to be considered since it requires per signal-dependent procedures. In [6], the authors recommended recursive thresholds for met data prepossessing and the authors highlighted that the proposed technique has weaknesses when it comes to detection of bias and drift, and it is successful for detecting well only spike errors and noise. Then, in [7], the authors published an analysis study based on the anemometers and wind vanes measurements with two different algorithms to develop an anomaly detection procedure using a manual check and proposed solutions using auxiliary variables such as wind speed ratio, wind speed Pearson

[1] Shared first authorship

correlations, and wind direction difference. Then they applied specific thresholds for anomaly labelling. Therefore, the authors have defined some rule sets based on their auxiliary variables and realised a model tuning for threshold selection. Whereas, in this paper, the rules are defined by machine learning algorithms, which use different measurement signals and do not limit the prepossessing analysis with a single threshold value or correlation analysis.

To the best of our knowledge, such a procedure is not developed and applied to meteorological data preprocessing up until this study. In normal conditions, such service requires minimum 90 minutes of an expert for a revision of a 1-month data, nevertheless via the proposed classifiers, this time need can be reduced to a couple of seconds.

2. Problem Definition

The data might inherit faulty registrations, the predefined rules are capable of detecting some part of these anomalies, but sometimes these rules become very restrictive and decrease the data availability without pointing to real anomalous status. Moreover, sometimes, although it is known that the sensor is not running and there is a noise in the signal, the predefined rules are not capable of labelling such anomalous situations. In Figures 1 and 2, some examples of the predefined rules-based filtering and when the experts are in agreement or disagreement with these rules are given.

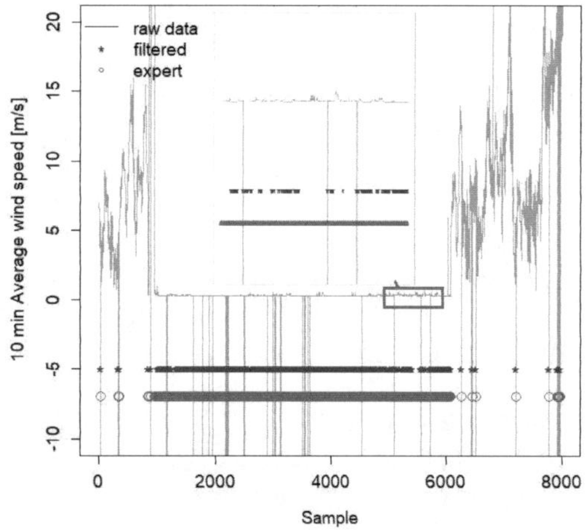

Figure 1. The difference between automatically filtered data and the expert reviewed data example 1.

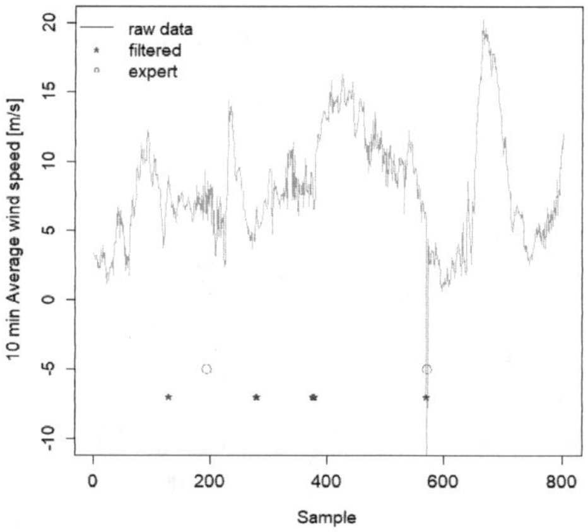

Figure 2. The difference between automatically filtered data and the expert reviewed data example 2.

Figure 1 shows a wind speed sensor anomaly. The anemometer is broken, therefore there are no measurement registries, but the recorder signal noise is registered. That is why the automatic fault detection programs do not detect the whole period as a sensor error. The detection of such error requires expert review and indication as an invalid period.

In Figure 2, the effect is the opposite, the automatic rule-based program detects faults that do not exist, and the expert made the review and retrieve the data identified as faults. This methodology involves time-consuming work by an expert and also, requiring significant manual intervention, increases the subjectivity of the decisions, leaving the validity of the data to the expert's judgement.

3. Data & Methodology

The input data consist of average, max, min, and standard deviation of wind speed, wind direction, temperature, and relative humidity data in 10 minutes resolution collected from different heights and sensor positions as listed in Table 1.

Table 1. Studied signals information

Number of signals	Sensor positions
4 Wind Speed(WS)	3 heights and 2 orientation
2 Direction Speed(WD)	2 heights
2 Temperature (T)	2 heights
2 Relative humedity (H)	2 heights

The proposed framework requires the availability of long term data, which consist of all possible labels and anomaly flags reviewed by experts. Then using clean datasets as training data, machine learning studies are performed.

3.1. Preprocessing common steps

Two input files not filtered data and filtered by experts are used as input. Data are divided into two subsets as flag columns and numerical measurement columns. Flag columns are defined as character columns. Not available data of numeric columns are filled using the previous valid value [8, 9]. Then for each wind speed measurement, the frequency of different flags are calculated in order to select a representative signal.

In 6 month learning period case, data start from 1/6/2019 and end 31/12/2019 are used. The remaining observations, which cover the period starts from 1/1/2020 to 31/8/2021 used as test data. In 9 month learning period case, data start from 1/6/2019 and end 31/03/2020 are used. The remaining observations, which cover the period starts from 1/4/2020 to 31/8/2021 used as test data. In 12 month learning period case, data start from 1/6/2019 and end 1/6/2020 are used. The remaining observations, which cover the period starts from 2/6/2020 to 31/8/2021 used as test data.

These three learning periods are studied in order to compare the effect of learning period in the classifier generation and availability of different flags.

3.2. Preprocessing for Wind model

The studied signal is from a sensor, which gathers the maximum wind speed data and is located at 60 meters height, 190 degrees, 2.1 m distance from the tower structure. The flags associated with this signal are the only character column.

The numerical columns consist of average wind speed, minimum wind speed, maximum wind speed, standard deviation measurements from the sensor located at 80 m height-10 degree, from the sensor located at 80 m height-190 degrees, from the sensor located at 60 m height-190 degree and from the sensor located at 40 m height-190 degrees.

3.3. Preprocessing for Wind-Hum-Temp model

The studied signal is from a sensor, which gathers the average wind speed data and is located at 80 meters height, 10 degrees, 2.1 m distance from the tower structure. The flags associated with this signal are the only character column. The numerical columns consist of the signals listed in the previous model and average, minimum, maximum and standard deviation measurements of temperature and relative humidity from the sensors located at 10 m height and 78 m height.

3.4. Preprocessing for Wind Direction model 1

The studied signal is from a sensor, which gathers the average wind direction data and is located at 78 meters height, 190 degrees, 2.1 m distance from the tower structure. The flags associated with this signal are the only character column. The numerical columns consist of the signals of average, maximum, minimum and standard deviation of wind direction measurements gathered from the sensors located at 78 and 58 meters and in the orientation of 190 degrees.

3.5. Preprocessing for Wind Direction model 2

This model uses the signals explained in Preprocessing for Wind Direction model 1. The only difference, here there is an additional data preprocessing step. This step converts numerical wind direction signals from degree to rad then apply a cosine function.

3.6. Machine learning models

In this study, 14 machine learning models have been tested by using the pycaret library, python programming language [10, 11], and the metrics accuracy, recall, and precision [12]. Please, see Table 2.

Table 2. Machine Learning Classifiers

Model	Abbreviation
Ada Boost Classifier	ada
Decision Tree Classifier	dt
Dummy Classifier	dummy
Extra Trees Classifier	et
Gradient Boosting Classifier	gbc
K Neighbors Classifier	knn
Light Gradient Boosting Machine	lightgbm
Linear Discriminant Analysis	Ida
Logistic Regression	Ir
Naive Bayes	nb
Quadratic Discriminant Analysis	qda
Random Forest Classifier	rf
Ridge Classifier	ridge
SVM - Linear Kernel	svm

3.6.1. Evaluation metrics: The models are compared considering three common performance metrics as accuracy, precision and recall [13]. These metrics can be calculated using the confusion matrix. The example of a binary classification problem will have a confusion matrix with four features [14]:

(i) True Positive (TP): The positive observations are labelled as positive (correct labels).
(ii) True Negative (TN): When the negative observations are labelled as negative (correct labels).
(iii) False Positive (FP): Negative observations are mislabelled as positive (misclassification).
(iv) False Negative (FP): Positive observations are mislabelled as negative (misclassification).

3.6.2. Accuracy

$$Accuracy = \frac{TP + TN}{TP + TN + FP + FN} \tag{1}$$

3.6.3. Precision

$$Precision = \frac{TP}{TP + FP} \tag{2}$$

3.6.4. Recall

$$Recall = \frac{TP}{TP + FN} \tag{3}$$

4. Results

In this section, the obtained results are presented for four models: Wind, Wind-Temp-Rel, Wind Direction 1 and Wind Direction 2. These models are defined considering flags of 3 signals: Maximum wind speed, average wind speed, average wind direction. For each model, all the classifiers are generated using 3 different learning periods as 6 month, 9 month and 12 month. In order to validate the best performing classifier, test data ([2/6/2020:31/8/2021], not used in the training phase) are used and generated anomaly estimations. Then, these anomalous labelled data are compared with auto filtered, and expert-reviewed data. In this study, expert reviewed data are assumed as the correct labelled data.

4.1. Wind model

In wind model case, dt, et, rf, knn models are found to be good performing models. Three applications of different learning periods are in agreement for this finding, see Figures 3,4,5.

The test sample shows us auto filtering cause, in this case, elimination of good samples, whereas machines learning generates very similar labelling to the expert review in Figure 6. Here Extra Trees (et) classifier is used.

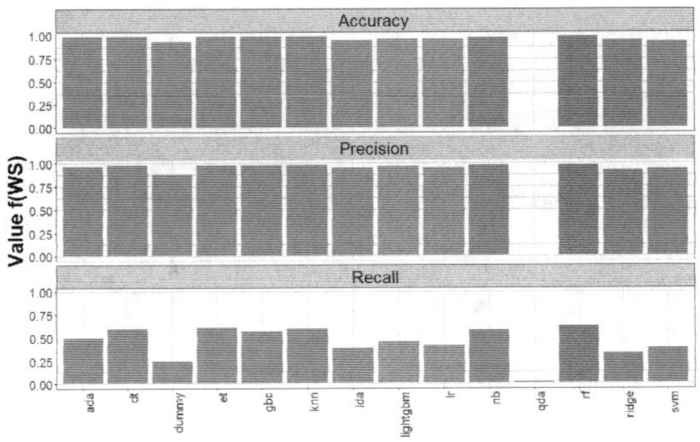

Figure 3. Wind Speed Model with 6 months learning period.

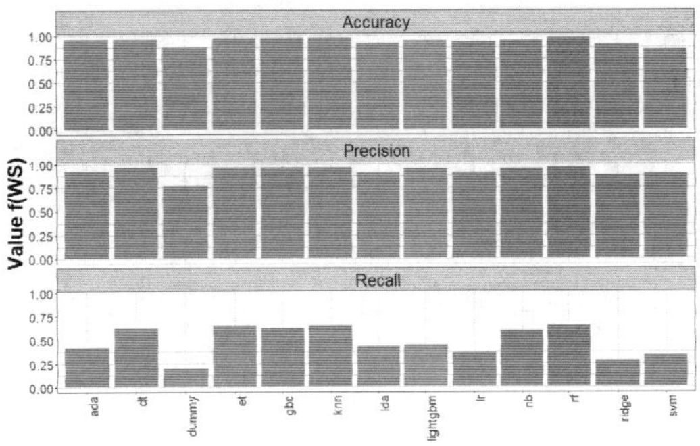

Figure 4. Wind Speed Model with 9 months learning period.

Journal of Physics: Conference Series 2257 (2022) 012015 doi:10.1088/1742-6596/2257/1/012015

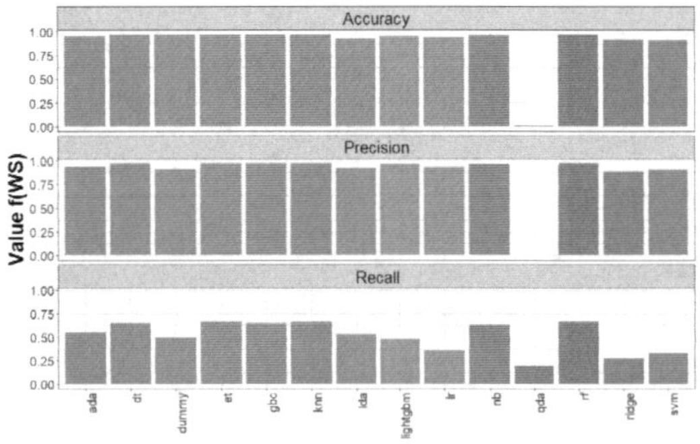

Figure 5. Wind Speed Model with 12 months learning period.

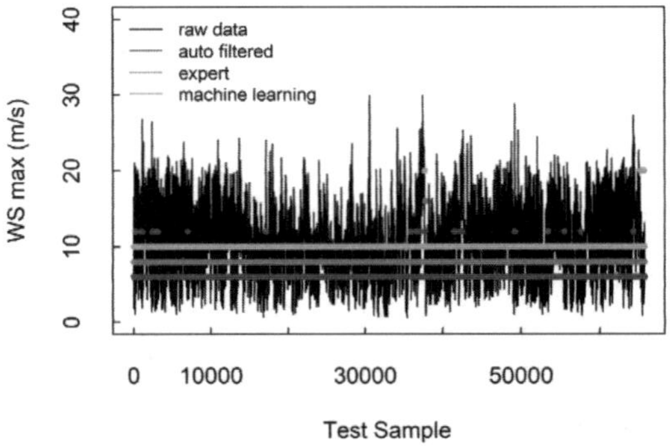

Figure 6. Sample Test Wind Model using the best performing classifier.

4.2. Wind-Temp-Rel model

In Wind-Temp-Rel model case, the effect of introducing more signals is studied. In this case, the best performing model changes depending on the learning period length, see Figures 7, 8, 9.

- 6 months learning period: nb is the best model
- 9 months learning period: dt, rf, nb, gbc are the best performing ones.
- 12 month learning period: dt is the best model

The test sample shows us auto filtering eliminates less data in comparison to expert review. In this model, machine learning labels more samples than the expert review 10. Here Decision Trees (dt) classifier is used.

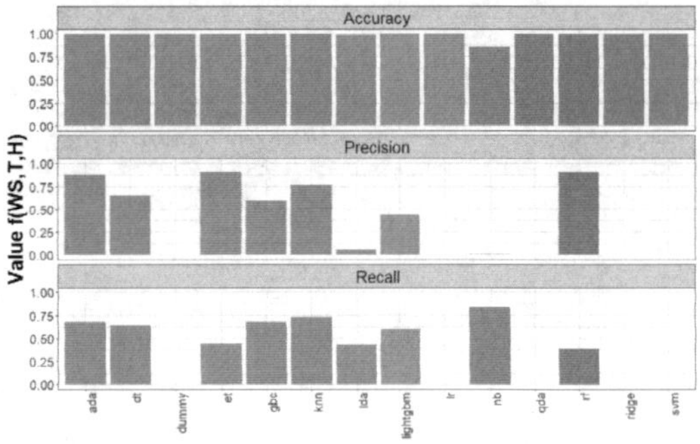

Figure 7. Wind-Temp-Rel Model with 6 months learning period.

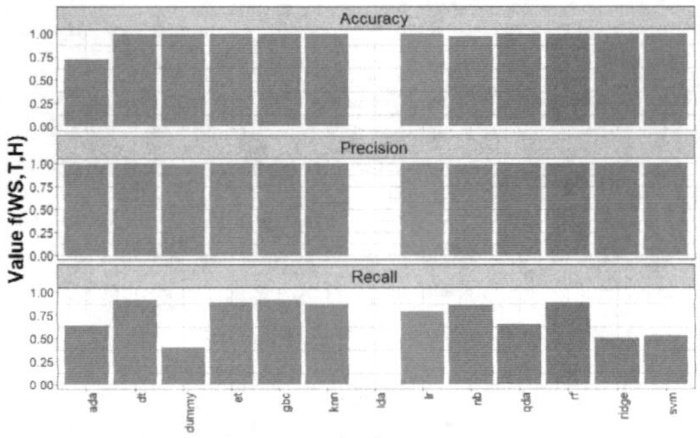

Figure 8. Wind-Temp-Rel Model with 9 months learning period.

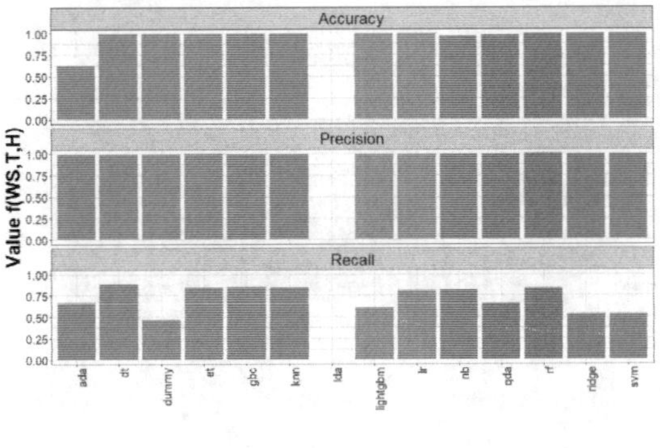

Figure 9. Wind-Temp-Rel Model with 12 months learning period.

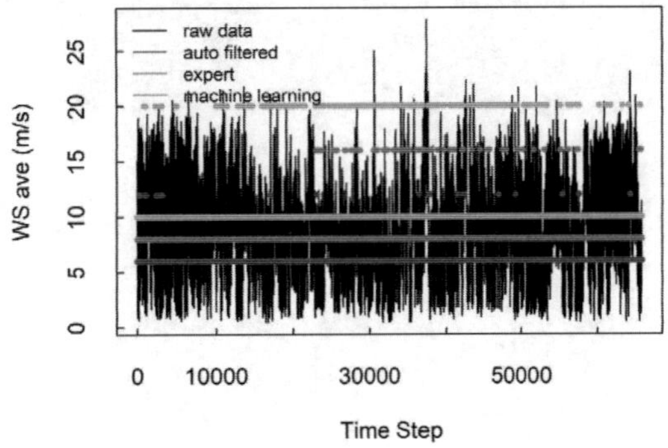

Figure 10. Sample Test Wind-Temp-Rel Model using the best performing classifier.

4.3. Wind Direction model 1

In Wind Direction model 1 case, the best performing model changes depending on the learning period length, see Figures 11, 12, 13.

- 6 months learning period: ada, et, gb, lr, nb and rf are the best performing ones.
- 9 months learning period: ada, dt, et and rf are the best performing ones.
- 12 month learning period: ada, et, knn, nb and rf are the best performing ones.

The test sample shows us that due to limited anomalous data observations, the machine learning classifier did not generate any anomalous label, whereas expert and auto filtering rules are in agreement for a very specific error, see Figure 14. Here Random Forest (rf) classifier is used.

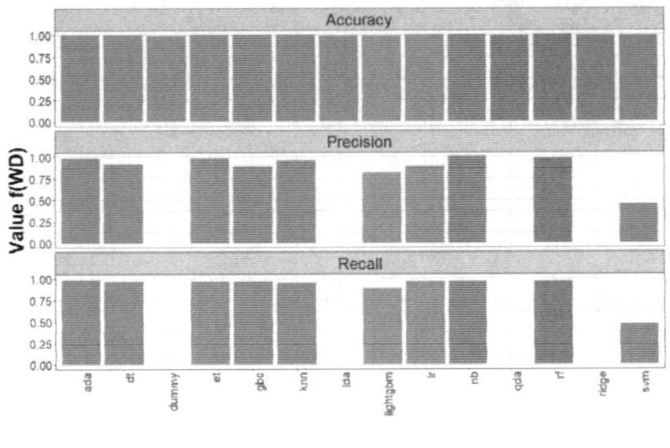

Figure 11. Wind Direction 1 Model with 6 months learning period.

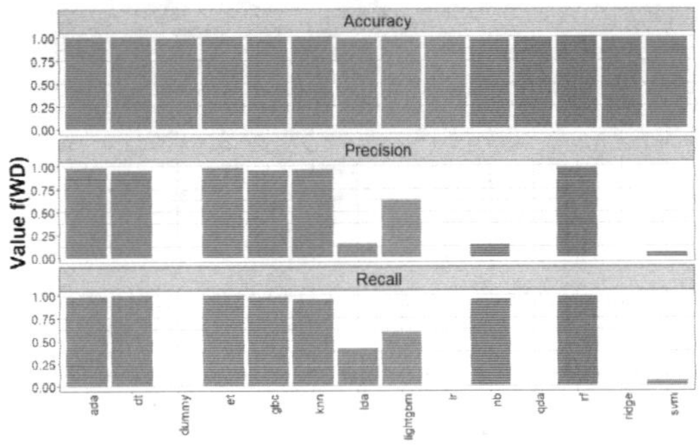

Figure 12. Wind Direction 1 Model with 9 months learning period.

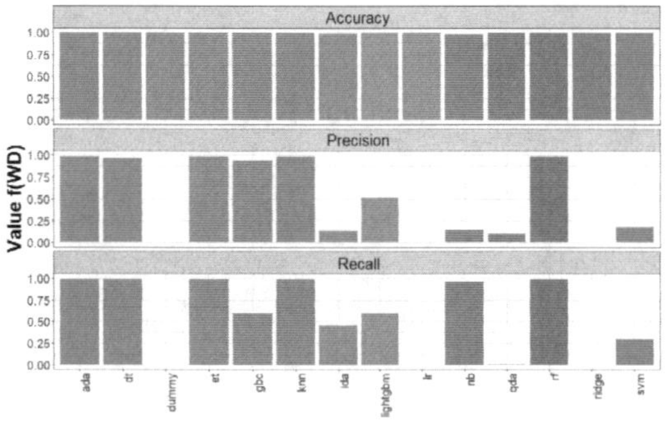

Figure 13. Wind Direction 1 with 12 months learning period.

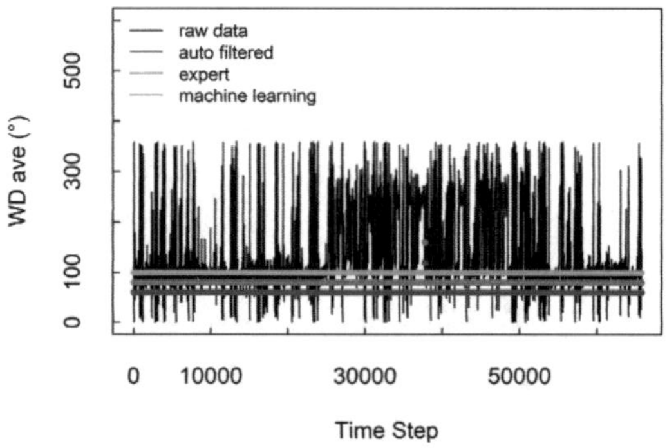

Figure 14. Sample Test Wind Direction 1 Model using the best performing classifier.

4.4. Wind Direction model 2

In order to see the influence of an additional data-preprocessing application, wind direction model 2 is generated. In Wind Direction model 2 case, the best performing model changes depending on the learning period length, see Figures 15, 16, 17.

Although the classifier is changed to ADA and the data treated differently, the result was the same with the Wind Direction Model 1 in the test sample evaluation see Figure 18.

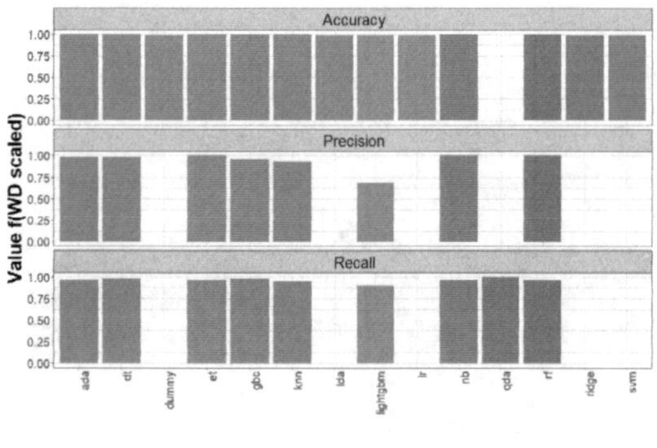

Figure 15. Wind Direction 2 Model with 6 months learning period.

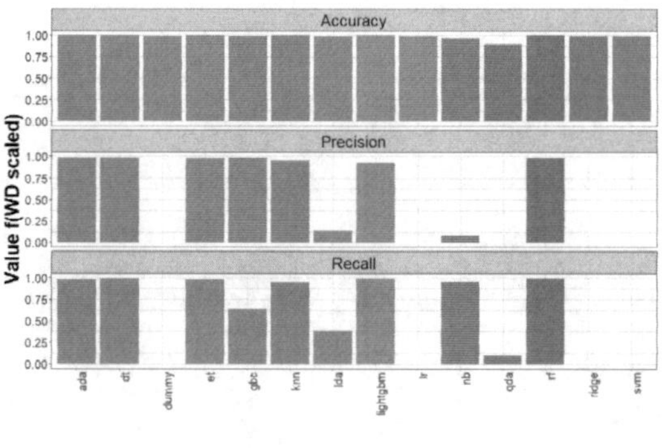

Figure 16. Wind Direction 2 Model with 9 months learning period.

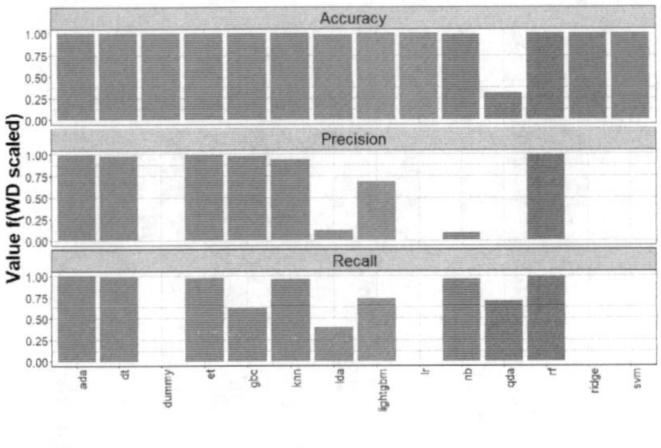

Figure 17. Wind Direction 2 Model with 12 months learning period.

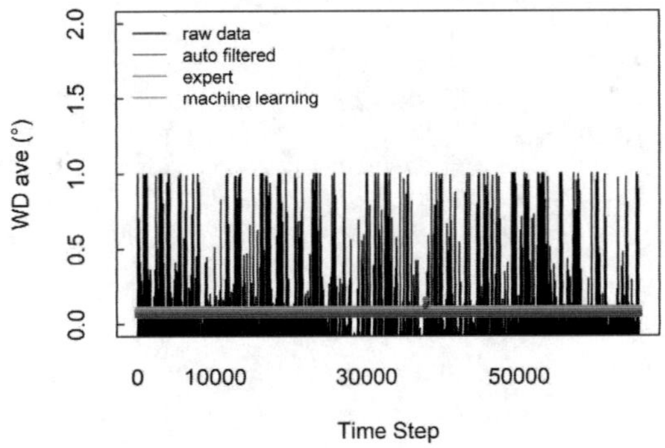

Figure 18. Sample Test Wind Direction 2 Model using the best performing classifier.

5. Conclusions

In this study, the improvement possibilities using machine learning for data filtering services are presented. Instead of using a single machine learning technique, it is recommended to apply various machine learning classifiers, since it is shown that when the analysed signal is changed, the best performing classifier is also changed. It must be highlighted that having a representative number of anomalies to be flagged in training data have utmost importance to generate successful classifiers. One of the significant findings is that the tree-based classifiers are performing better in comparison to other machine learning models. By benefitting from the classifiers generated by machine learning algorithms preprocessing phase of the wind resource assessment project can be shortened.

Acknowledgments

The authors would like to acknowledge the financial support used in the investigation of machine learning applications for renewable energy generation data in the project ENERISLA financed by RED CERVERA program.

References

[1] IEC Wind energy generation systems – Part 12-1: Power performance measurements of electricity producing wind turbines
[2] Vaumi J P T, Yenke B O, Bame N and Sarr I 2018 *2018 14th International Conference on Signal-Image Technology & Internet-Based Systems (SITIS)* (IEEE) pp 574–579
[3] Stetco A, Dinmohammadi F, Zhao X, Robu V, Flynn D, Barnes M, Keane J and Nenadic G 2019 *Renewable energy* **133** 620–635
[4] Liu H and Chen C 2019 *Applied Energy* **249** 392–408
[5] Beltrán J Llombart A G J 2009 *International Conference on Renewable Energies and Power Quality (ICREPQ'09)*
[6] Hasu V, Koivo H and Poutiainen J 2006 *First International Conference on Innovative Computing, Information and Control-Volume I (ICICIC'06)* vol 3 (IEEE) pp 140–144
[7] Guida F P, Iaconis M and Visco-Comandini F 2020 *2020 IEEE Congreso Bienal de Argentina (ARGENCON)* (IEEE) pp 1–5
[8] Wickham H, Averick M, Bryan J, Chang W, McGowan L D, François R, Grolemund G, Hayes A, Henry L, Hester J, Kuhn M, Pedersen T L, Miller E, Bache S M, Müller K, Ooms J, Robinson D, Seidel D P, Spinu V, Takahashi K, Vaughan D, Wilke C, Woo K and Yutani H 2019 *Journal of Open Source Software* **4** 1686

[9] R Core Team 2021 R: A language and environment for statistical computing URL https://www.R-project.org/
[10] Python Software Foundation Python Language Reference, version 3.9. (accessed: 01.10.2021) URL http://www.python.org
[11] Ali M 2020 Pycaret: An open source, low-code machine learning library in python pyCaret version 1.0.0 URL https://www.pycaret.org
[12] Joseph Rickert Evaluation metrics (accessed: 01.10.2021) URL https://blog.revolutionanalytics.com/2016/03/com _class_eval_metrics_r.html
[13] Campesato O 2020 *Artificial Intelligence, Machine Learning, and Deep Learning* (Stylus Publishing, LLC)
[14] Ceja E G 2021 *Behavior Analysis with Machine Learning Using R* (Chapman and Hall/CRC)

AUTHOR INDEX